400余种手绘多肉植物图谱
400 Colour Paintings of Succulent Plants

朱亮锋　编著

http://www.hustp.com

中国·武汉

图书在版编目（CIP）数据

400余种手绘多肉植物图谱/朱亮锋编著. -- 武汉：华中科技大学出版社，2017.10
ISBN 978-7-5680-3410-4

Ⅰ.①4… Ⅱ.①朱… Ⅲ.①多浆植物－图谱 Ⅳ.①S682.33-64

中国版本图书馆CIP数据核字（2017）第236965号

400余种手绘多肉植物图谱
400 Yuzhong Shouhui Duorou Tupu

朱亮锋 编著

出版发行：华中科技大学出版社（中国·武汉） 电话：（027）81321913
地　　址：武汉市东湖新技术开发区华工科技园（邮编：430223）
出 版 人：阮海洪

策划编辑：王　斌	责任监印：朱　玢
责任编辑：吴文静	装帧设计：百彤文化
校　　对：李用华	

印　　刷：雅昌文化（集团）有限公司
开　　本：965mm×1270mm　1/16
印　　张：14.5
字　　数：300千字
版　　次：2017年10月第1版　第1次印刷
定　　价：188.00元（USD 37.99）

投稿热线：（020）66636689　　342855430@qq.com
本书若有印装质量问题，请向出版社营销中心调换
全国免费服务热线：400-6679-118 竭诚为您服务
版权所有　侵权必究

朱亮锋，教授，研究员，汉族，1937年1月出生于广州，1960年毕业于中山大学化学系，1964年在中国科学院华南植物研究所从事植物化学和植物资源学的研究，曾任华南植物研究所植物资源研究室副主任、主任，从1987年开始从事芦荟和中药材研究开发工作，曾任联合国教科文组织（UNESCO）亚洲及太平洋地区药用植物和芳香植物情报网络（APINMAP）联络员和中国国家中心负责人。1989年获国务院有突出贡献科学家津贴；1999年获国际精油和香料行业联盟（IFEAT）成就奖章。1999年退休。

Zhu Liangfeng, Professor, born on January 15, 1937 in Guangzhou, who graduated from the Department of Chemistry of Zhongshan University in 1960. He engaged in Phytochemistry and Plant Resource research at South China Institute of Botany, Academia Sinica from 1964 and started the research and development works of Aloe from 1987. He used to be the head of the department of plant resources, the liaison officer and the head of Chinese National Center of UNESCO Asian Pacific Information Network on Medicinal and Aromatic Plants (APINMAP). He also gained the Notable Contribution Scientist Allowance from State Council of the People's Republic of China in 1989 and the Achievements Medal from International Federation of Essential Oils and Aroma Trades (IFEAT) in 1999. Recently, he is the member of technical committee and the specially invited director of the Aloe Society of China.

吴萍，女，汉族，博士，1969年10月生于遵义，1992年毕业于兰州大学生物系，现为中国科学院华南植物园副研究员。

Wu ping Ph.Dr. born in Zunyi City, October 1969. Graduated from Department of Biology, Lanzhou University in 1992, Currently assistant Professor of South China Botanical Garden, Chinese Academy of Sciences.

王辰，男，汉族，1988年4月出生于兰州。2010年毕业于兰州大学化学化工学院，现于中国科学院华南植物园攻读博士学位。

Wang Chen, male, born in Lanzhou, April, 1988. Graduated from Department of Chemistry and Chemical Engineering, Lanzhou University in 2010. Currently as a Ph.D. candidate of South China Botanical Garden, Chinese Academy of Sciences.

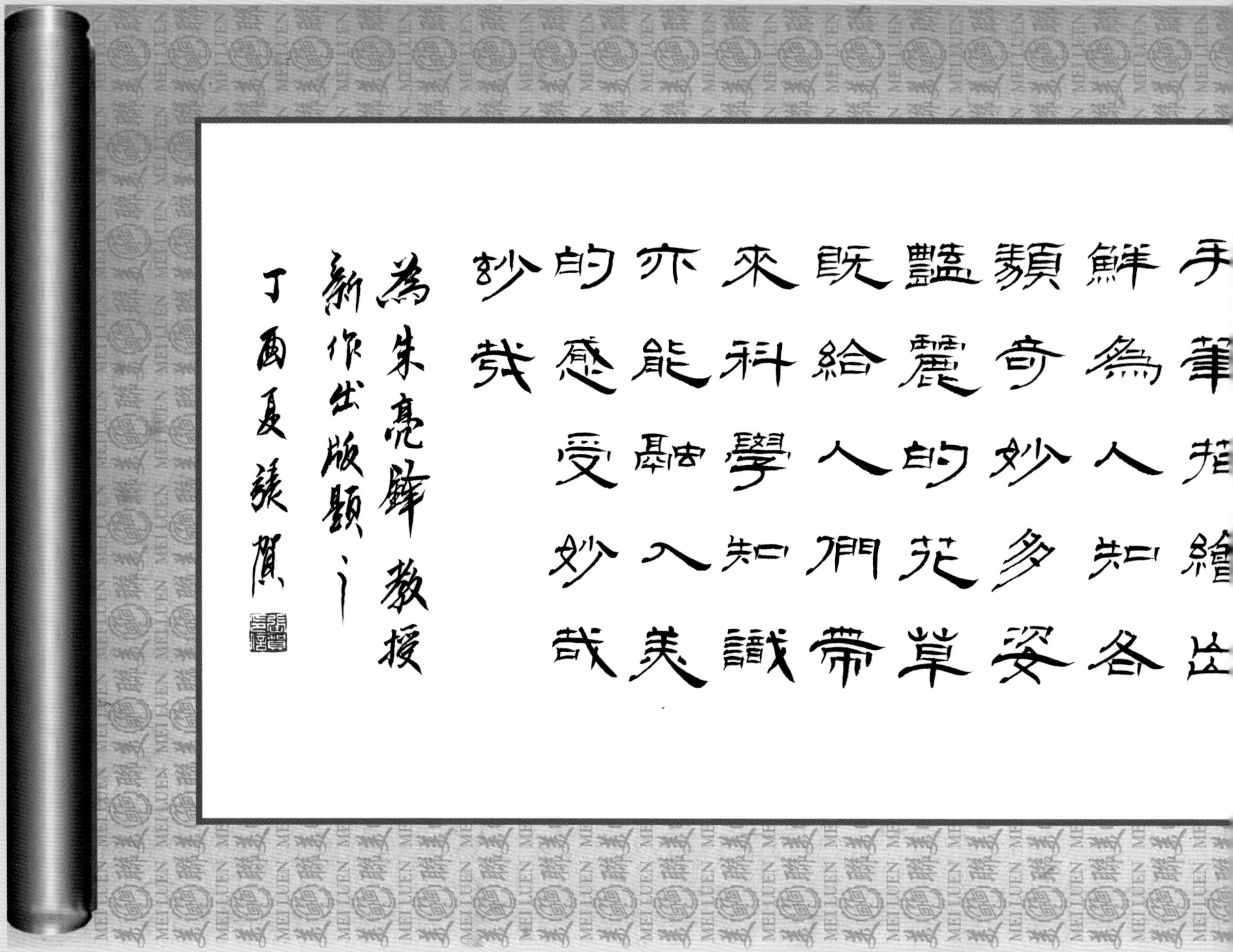
手筆描繪出
鮮為人知各
類奇妙多姿
豔麗的花草
既給人們帶
來科學知識
亦能融入美
的感受妙哉
妙哉
為朱亮鋒教授
新作出版題之
丁酉夏張賀

大年生按自規，津定事于人之，然科學的之，踏上暮年繪肉書，時用手一沙，質植物沙，投身于雲海之，藝術奇我析，中奇我，運用藝術。

前言

50多年的爱好和40多年的收集和积累，在各位同窗好友的帮助、鼓励和支持下，终于在退休后年龄接近80岁的途中，如愿将400余种多肉植物彩色图谱呈现给大家。

在退休前借工作之便，偷闲到世界各地多肉植物的主要分布地区和各地区的植物园、植物公园拍照收集，如，有"仙人掌王国"之称的墨西哥、美国东南部多个州的国家公园和自然分布地区、加勒比海一些仙人掌分布岛屿等；百合科芦荟属、番杏科部分种属主要分布地区：南非、比勒陀尼亚、斯威士兰以及它们的植物园。另外还到过南半球的新西兰威灵顿植物园、澳大利亚墨尔本植物园和悉尼植物园、东南亚印尼的茂物和巴厘岛植物园、新加坡植物园、法国巴黎植物园、摩纳哥植物园、加拿大温哥华室内植物园等。当年也走遍了我国各大植物园：北京香山植物园、南京中山植物园、昆明植物园、西双版纳植物园、厦门植物园、深圳仙湖植物园、武汉植物园、庐山植物园、西安植物园、贵阳植物园、上海龙华植物园和华南植物园等，前后共收集四千多张照片和有关资料，再经多年整理、筛选、汇总和查阅相关资料，最后临摹成画与大家见面。

既然是手绘的画，主要是将原始素材的每一种多肉植物通过照片、图片临摹加工成画，由于作者并非专业绘画之人，也并非植物分类学家、园艺栽培专家，对于确定每种多肉植物的命名和形态分布描述，在多名专家指点下还算见得人，但我估计错漏还是免不了的，敬请大家多多包涵。在本图谱中，由于时间有限不能把一些栽培种的形态和出处加以介绍。

400余种手绘多肉植物图谱是通过多人参与，照片整理、资料收集、中英文简介、临摹过程的指点、每种图的扫描、出版等环节，少一个都不能完成，所以这本图谱不只是作者本人的，而是所有参与者的，在此对他们表示谢意。

PROLOGUE

Before my retirement, by taking the advantage of business trip, I had visited many botanical gardens, botanical parks, and natural distribution areas of succulent plant all around the world to collect photographs and pictures, including Mexico- "the cactus kingdom"; natural distribution areas and national parks in southeast America; several islands of Caribbean Sea; South Africa and Swaziland, where are the major distribution areas of genus Aloe and several genera and species of family Aizoaceae; Wellington Botanical Garden in New Zealand; Melbourne and Sydney Botanical Gardens in Australia; Kebon Raya Bogor and Bali Botanical Garden; Singapore Botanical Garden; Paris Botanical Garden in France; Monaco Botanical Garden; Vancouver Botanical Garden in Canada; and most of the botanical gardens in China, such as Xiangshan Botanical Garden in Beijing, Zhongshan Botanical Garden in Nanjing, Kuming Botanical Garden, Xishuangbanna Botanical Garden, Xiamen Botanical Garden, Xianhu Botanical Garden in Shenzhen, Wuhan Botanical Garden, Lushan Botanical Garden, Xian Botanical Garden, Guiyang Botanical Garden, Longhua Botanical Garden in Shanghai and South China Botanical Garden in Guangzhou. Over 4000 photographs had been taken during these trips, and after years of selection, organization, summary and review, I have finally finished these drawings.

Since they are hand painted, all of the drawings are copies of the original photographs or pictures. People who have involved in the production and publication process of these illustrations, including myself, are neither professional painters, horticulture experts, nor phytotaxonomists. All the captions are simple, rough introductions of the specified illustrations, and you may find some inappropriate expressions between the lines. I sincerely wish that you readers could forgive my mistakes in the nomenclature, morphology, and distribution descriptions of these succulent plants.

Without the help from my dear friends and colleagues, these hand painted drawings of over 400 succulent plant species would not have been finished. Hereby I deeply thank the ones who have helped me with the photograph organization, data collection, Chinese and English introduction, painting direction, pictures scan, and publishing procedure.

目录 CONTENTS

仙人掌科 (Cactaceae)

花冠球 /3
明美玉（阿寒玉）/3
龙舌兰牡丹锦 /4
黄体象牙牡丹锦 /4
猩猩冠柱 /5
格氏关节仙人掌（节状仙人柱）/5
多头兜 /6
琉璃兜锦 /6
黄兜锦 /7
鸾凤兜 /7
瑞凤玉 /8
五棱碧鸾凤玉 /8
鸾凤玉（僧帽）/9
四棱鸾凤玉 /9
四角碧鸾凤玉 /10
六棱鸾凤玉 /10
碧琉璃鸾凤玉 /11
七棱鸾凤阁 /11
白云般若 /12
白云般若 /13
群凤玉 /13
金色火炬（天鹅绒仙人柱、碧彩柱、丝绒柱）（碧彩柱属）/14
伟冠柱 /15
雨树花冠柱 /15
雨树花冠柱 /16
丝毛花冠柱 /16
佛塔 /17
未鉴定仙人掌科植物 /17
钢盔天轮柱 /18
姬白云狮子锦 /18
鬼面阁 /19
山吹 /19
鲜丽玉锦 /20
阿根廷银毛柱 /20

烛焰管花柱（里特里管花柱）/21
黑士冠 /21
百花龙爪玉 /22
地中生龙爪玉 /22
毛疣仙人球 /23
鱼鳞玉 /23
鱼鳞丸 /24
黑王玉 /24
金黄恐龙角 /25
菠萝球（天司球）/26
喜钙顶花球 /26
狮子奋迅 /27
海胆仙人掌 /27
白豹（纤细顶花球）/28
劳氏顶花球 /28
枪骑士（长须顶花球）/29
圣克鲁斯蜂窝仙人球（顶花球属）/29
粗刺顶花球 /30
菠萝（球）仙人掌（顶花球属）/30
萨迪顶花球 /31
具槽顶花球 /31
花精丸 /32
隐柱天轮柱 /32
隐柱天轮柱（分叉）/33
火焰球（变种）/33
火焰球 /34
茜球 /34
七刺圆盘玉 /35
赫氏圆盘玉 /35
碟形圆盘玉 /36
尼逊双重仙人掌 /36
红花双重仙人掌 /37
金鯱 /37
裸鯱 /38
短刺金鯱 /38
花王丸 /39
巨金鯱 /39

岩（广利球）/40
狂刺凌波 /40
粗刺凌波 /41
凌波 /41
紫红玉 /42
巴氏鹿角柱 /43
金龙 /43
赤寿山鹿角柱 /44
司虾（幻虾、武勇丸）/45
御旗 /45
尼克鹿角柱 /46
九刺虾 /46
九刺虾杂交变种 /47
卫美玉 /47
富丽鹿角柱 /48
幻虾 /48
幻虾（变种）/49
明矾 /49
大花鹿角柱杂交变种 /50
钩刺鱿 /50
宇宙殿 /51
劳氏鹿角柱 /51
海胆（魔剑丸、豪剑丸）/52
王将虾 /52
墨西哥微刺虾（变种）/53
春高楼 /53
柏玛尼逊鹿角柱 /54
三光球 /55
黄太阳 /55
刺虾 /56
明石丸 /56
姬路丸 /57
平和虾 /57
花杯 /58
锦照虾（折墨虾）/59
珠毛柱 /59
鲜红虾（富丽虾）/60

大佛殿 /60
阿巴尼斯虾 /61
月光虾 /61
荒武者（草虾、集合虾）/62
奥克特大佛殿 /62
三刺虾杂交变种 /63
青花虾杂交变种 /63
莫加维皇杯（变种）/64
鲜红鹿角柱 /64
美桃虾 /65
青花虾 /65
戴维青花虾 /66
鬼见城 /66
瓦诺克美刺球 /67
鹿角柱属杂交种 /67
弯刺仙人球 /68
芳春丸 /68
金城柱 /69
卡氏弯刺仙人球 /69
世界图（地图宝）/70
短刺仙人球 /70
黑刺短毛丸 /71
赤盛丸 /71
仙人球属杂交种 /72
绯盛丸 /72
鲜凤丸（红凤丸、鸡冠掌）/73
稀刺仙人球 /73
十棱仙人球 /74
红百合仙人球 /74
仙人球属一种 /75
白弯刺仙人球 /75
托氏仙人球 /76
仙人球属杂交种 /76
黄心昙花 /77
黑闪光（呵云玉）/78
黑闪光 /78
李氏须弥丸 /79

紫王子 /79
玫瑰红松球 /80
横纲 /80
秘鲁幻乐 /81
白丽翁（立翁柱）/81
鹰之巢 /82
白银城 /82
龙王球 /83
黄花强刺球 /83
狂刺江守 /84
龙虎 /84
黄绿强刺球 /85
福氏强刺球 /85
直刺强刺球 /86
弯刺强刺球 /86
大虹 /87
大蜓 /87
文鸟 /88
金鸥玉 /88
天城 /89
黄彩玉 /89
毛花筒球 /90
鳄之子 /90
姬毛玉 /91
圣王球 /91
绯牡丹锦 /92
新天地锦 /92
白花裸萼球 /93
分生裸萼球 /93
绯花玉（瑞昌玉）/94
黑枪球 /94
宝卵球 /95
碧岩玉 /96
黄蛇丸 /97
翠皇冠 /97
绯花玉（变种）/98
快龙丸 /98

罗星丸 /99
布氏裸萼球 /99
粉冠丸 /100
桃冠球 /100
光琳玉 /101
良宽 /101
红花天王球杂交种 /102
丽蛇丸 /102
蛇龙丸 /103
天王球杂交种 /103
长刺勇将球 /104
海王丸 /104
九纹龙 /105
翠皇冠 /105
菲车士裸萼球 /106
荷氏裸萼球（圣王球）/106
雪冠球 /107
丛生裸萼球 /107
多花牡丹玉锦 /108
牡丹玉锦 /108
瑞云牡丹锦 /109
碧严玉 /109
黑罗汉丸（紫冠玉、拉根）/110
武碧玉 /110
云龙（观音龙、纹美玉、纹绯丽）/111
波光球（华龙丸、琥龙丸、琥龙光、新天龙、龙光丸、龙华丸、爬荒龙、蛇斑龙）/111
须黑玉 /112
土蜘蛛 /112
豪刺花王丸 /113
青玉丸 /114
天平丸 /115
绫鼓（碧盘玉、翠盘玉）/115
绫鼓（碧盘玉、翠盘玉）/116
内弯刺裸萼球 /116
守金摩天龙 /117
恐龙球（钟鬼丸）/117
巨锁龙（月亮柱、新桥）/118
长钩玉 /118
砂王女（伊须萝玉）/119
光虹球 /119
黄裳丸 /120
豪剑丸 /121

湘阳丸 /121
梦春丸 /122
粉花芳春丸 /122
龟甲丸 /123
红花阳盛球（阳盛丸）/124
红火炬仙人球 /124
太刀姬 /125
红笠丸（朱丽丸、朱丽球）/127
劳氏丽球 /128
丽刺丸 /129
凄厉丸 /129
优髯球 /130
花扇丸 /130
花生仙人球 /131
脂粉丸（无忧愁）/132
哲克仙人球 /133
乌羽玉 /133
翠冠玉 /134
翠冠玉（吹子）/134
翠冠玉 /135
地久丸 /135
明显乳突球 /136
白鹭 /136
樱富士（明日之光）/137
雪头球 /137
白豹丸 /138
吉赛尔 /138
红艳锦 /139
猩猩丸 /139
富贵丸 /140
舞衣 /140
海仙玉 /141
白玉仙 /141
普氏白仙玉 /142
丽云 /142
生长在火山岩上的花座球 /143
翠云 /144
小花仙人柱 /144
龙神木 /145
勇风 /145
黑冠丸（素妆玉）/146
大轮丸 /146
海氏激光球 /147
豹头 /147

雷头玉 /148
逆龙玉 /148
短刺智利球 /149
令箭荷花 /149
白乐天（白小町）/150
丛生南国玉 /150
海神丸 /151
棕色锦绣玉 /151
红彩玉 /152
短刺绒毛南国玉 /152
英冠玉 /153
妙梅锦绣玉 /153
青王球 /154
珊瑚城（白毯丸）/155
红刺南国玉 /155
灰刺南国玉 /156
眩美玉 /156
瓦拉斯锦绣玉 /157
黄花海王球（变种）/158
密花团扇 /158
使命仙人掌（金枪无花果、印度无花果）/159
黄花仙人掌 /159
白刺仙人掌（白毛掌）/160
单刺仙人掌 /160
棕色脊椎刺梨（仙人镜）/161
土人团扇 /161
红刺黄花仙人掌 /162
将军 /162
清淡仙人掌（银椢）/163
圣云龙 /163
武伦柱 /164
黄心锦绣玉 /164
锦绣玉 /165
黄刺锦绣玉 /165
柱状锦绣玉 /166
锦绣玉 /166
舞绣玉（快武丸）/167
暗绿锦绣玉 /167
丽花球 /168
郝氏锦绣玉 /168
胡梅尔锦绣玉 /169
卡拉锦绣玉 /169
劳氏锦绣玉 /170

马里锦绣玉 /170
梅赛德斯锦绣玉 /171
青王丸 /171
血红花锦绣玉 /172
绯绣玉 /172
绯宝球（变种）/173
红绣玉 /173
斯图海锦绣玉 /174
韦伯锦绣玉 /174
精巧丸 /175
精巧殿 /175
金凤龙 /176
巴西青毛柱 /176
舟乘团扇 /177
新玉（白银宝山）/177
眼球子孙球 /178
华斯子孙球 /178
绯宝球 /179
白花子孙沟宝球 /180
白花沟宝球 /180
白宫丸变种 /181
黑丽球 /182
茶柱 /183
有沟宝山球 /183
黑丽丸 /184
瓦斯奎斯有沟宝山球 /184
梦托莎 /185
轮刺有沟宝山球 /185
大统领 /186
博拉大统领 /186
瓦兹大统领 /187
赤岭丸 /187
天晃 /188
龙王丸 /188
具棱瘤玉球 /189
具棱瘤玉球变种 /189
金光龙 /190
黑凤 /190
利升龙 /191
金棱 /191
蛟丽丸 /192
精巧殿 /192
类节球 /193
金刺尤伯球 /193

黄刺尤伯球 /194
黄刺节齿尤伯球 /194
花细球 /195

百合科 (Liliaceae)

铃丽锦（皮刺芦荟）/197
非洲芦荟 /197
盎格鲁芦荟 /198
布尔芦荟 /198
绿花芦荟 /199
箭筒芦荟（二分叉芦荟）/199
津巴布韦芦荟（高芦荟）/200
格斯特芦荟 /200
鬼切芦荟 /201
巨箭筒芦荟 /201
雷诺兹芦荟 /202
超雪芦荟 /202
奇丽芦荟 /203
艳丽芦荟（歪头芦荟）/203
瓦奥姆比芦荟 /204
恐龙卧牛（短叶卧牛一种）栽培种 /204
霸龙 /205
卧牛锦 /205
比兰龙锦 /205
卧牛龙白覆轮 /206
白摺鹤 /207
点纹冬之星座 /207
万象锦 /208
玉扇锦 /208
鬼瓦 /209

景天科 (Crassulaceae)

黑法师 /211
妲星（黑法师变种）/212
新花月锦 /213
艳镜 /213

番杏科 (Aizoaceae)

慈光锦 /215
翔凤 /215
奇凤玉 /216
玉帝 /216

仙人掌科 (Cactaceae)

仙人掌科植物是一个非常庞大的家族，绝大多数种属为肉质植物，按前期的植物分类研究被归类为 233 个属、670 种，但尚未能把所有杂交、变种归纳进去。仙人掌科植物由于种与种之间、属与属之间不易划分，故还在不断归并和分类、变化不断，至 1993 年仙人掌科又被合并为 98 属，现在看来这个大家族还在不断演变中。

大多数仙人掌植物随着环境变化，其形态不断适应，最明显就是其叶在干旱恶劣环境中已经变成针刺状和长短不一的绒毛状，而它的茎则变成肥大肉质、环状、筒状、柱状等形态，使之形成与其他植物不同的外貌。仙人掌科千姿百态、变幻神奇的植株为世人所喜爱，故仙人掌已成为各类温室必备的盆栽观赏花卉。

Cactaceae is a huge family which could be subdivided into 233 genera and 670 species according to previous taxonomy studies. However, all of the hybrids and variants are not included. Due to the difficulty in dividing the plants of family Cactaceae between species and species, genera and genera, the merging and classification of them have not been finished yet. The Cactaceae has been merged into 98 genera in 1993, it seems that this huge family is still evolving.

Most of the Cactaceae plants could constantly adapt to the changes of environment. As a result of growing in drought conditions, their leaves are usually needle-like, stem fleshy, succulent, circular, cannular or columnar. Their diverse and particular morphologies are deeply favored by people, and make them become the indispensable potted ornamental flowers in greenhouse.

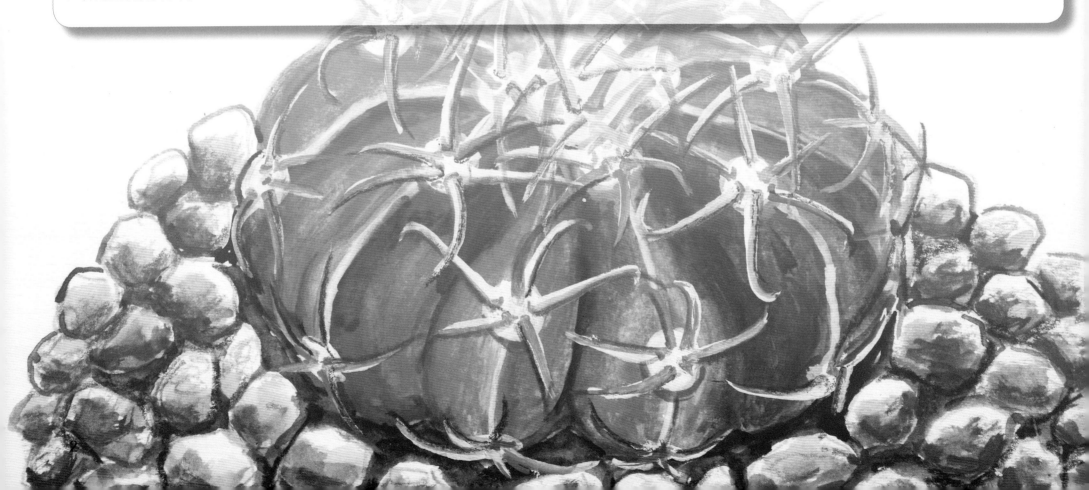

花冠球
Acanthocalycium thionanthum (Speg.) Bkbg.

植株单生，高达 12 cm，宽 6~10 cm。变种 var. *brevispinum* (Ritt.) Don 高达 50 cm，宽 6 cm，刺较原种更短、颜色更深，花黄色。原产阿根廷。

Solitary, up to 12 cm high and 6~10 cm wide. var. *brevispinum* (Ritt.) Don is up to 50 cm high and 6 cm wide, spines are darker and shorter than the original species, flowers yellow. Native to Argentina.

明美玉（阿寒玉）
Acanthocalycium thionanthum var.aurantiacum (Rausch) Don. ex Rausch

植株单生，高达 5 cm，9 cm 宽。5~7 根周刺，5 cm 长。1 根中刺。花直径 5 cm。变种 var. *ferrarii* 宽达 12 cm，9 根周刺，1~4 根中刺，2 cm 长。花红色。原产阿根廷 Catamarca。

Solitary, up to 5 cm high, 9 cm wide. 5 to 7 radial spines, 5 cm long. 1 central spine. Flowers 5 cm in diameter. var.ferrarii is up to 12 cm wide, 9 radial spines, 1 to 4 central spines, 2 cm long. Flowers red. Native to Catamarca, Argentina.

龙舌兰牡丹锦
Ariocarpus agavoides 'Variegata'

为龙舌兰牡丹的斑锦品种,茎扁平,株高 2~2.5 cm,株幅 6~8 cm,茎顶端簇生细长三角形疣突,表皮角质,初生深绿色,后逐渐转为灰绿色,基部橙黄色。刺座位于疣尖下 1 cm 处,有很厚的短绵毛,偶有 1~3 枚浅色短刺。花着生新刺座绒毛中,钟状花,玫瑰红色。

A variant of Ariocarpus agavoides, stems flat, 2 to 2.5 cm high, 6 to 8 cm wide. Slim, triangular tubercles grow densely on the tip of the stems. Body dark green when young, gradually trun into greyish green. Areoles locate under the tips of tubercles, covered with thick, short tomentum, 1 to 3 short spines occasionally appear on the areoles. Flowers are surrounded by the tomentum, bell-shaped, rosy in color.

黄体象牙牡丹锦
Ariocarpus furfuraceus 'Magnificus Variegata'

为花牡丹的一个斑锦品种。植株单生,株高 5~6 cm,株幅 10~12 cm,莲座状,疣突肥大饱满,宽三角形,全体黄色。花单生,钟状,白色或淡粉红色。秋季开花。

A variant of Ariocarpus furfuraceus. Solitary, body 5 to 6 cm tall, 10 to 12 cm wide, lotus~shaped, yellow in color. Tubercles are fleshy, broad trianglular in shape. Flowers solitary, bell~shaped, white or pale pink in color, appear in autumn.

猩猩冠柱
Arrpjadpa penicillata (Garke) Br. et R.

植株灌木状，易分枝，枝条生长时相互倚靠，最长可达 2 m。10~12 条低棱，周刺 8~12 枚，中刺 1~2 枚，为细针刺形。花顶生成簇，淡绯红色。原产巴西。

Shrub like, usually branched. Branches lein on each other, up to 2 m long. 10 to 12 ribs, 8 to 12 radial spines, 1 to 2 central spines, tiny needle~shaped. Flowers red, terminal and clustered. Native to Brazil.

格氏关节仙人掌（节状仙人柱）
Arthroereus glaziovii (K. Schum.) N.P. Taylor

植株平卧分枝或不分枝。茎节呈圆柱形或棒形，约有 12 条棱。周刺 25~35 根，白色，中刺 1~2 根。原产巴西。

Sometimes branched, sometimes not. Stems columnar or bar~shaped. 25 to 35white radial spines, 1 to 2 central spines. Native to Brazil.

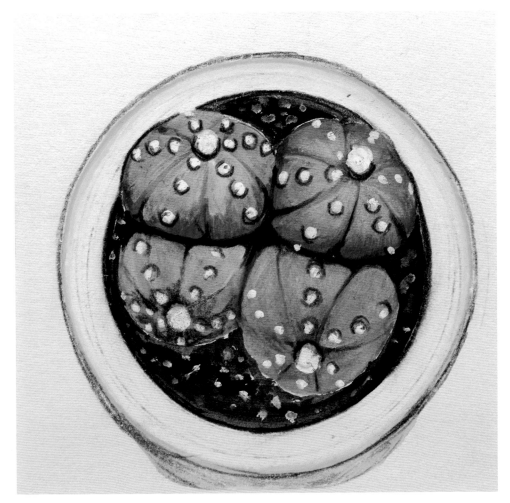

多头兜
Astrophytum asterias 'Caespitosa'

为兜的畸形变种。是一个由6~8个兜球组成的连体兜。每个球体为半球形，具8条棱，表皮灰绿色，星散分布由白色丛卷毛组成的星点，棱脊中央间隔长有绒毛状刺座。花顶生，漏斗状，黄色，花径3~7 cm，喉部红色。花期夏季。

A variant of A. asterias. 6 to 8 heads, semi-globular in shape. Body greyish green. 8 ribs with areoles grow btween each ribs. Flowers terminal, funnel~shaped, 3 to 7 cm in diameter, yellow in color with red throat, appear in summer.

琉璃兜锦
Astrophytum asterias 'Nudas Variegata'

为琉璃兜的斑锦品种。植株单生，扁球形，株高5~10 cm，株幅6~10 cm，具8个整齐宽棱，棱面青绿色，带黄色隐斑，无星点，棱脊中央着生纵向绒球状刺座。花漏斗状，淡黄色，花径3~4 cm。花期夏季。

A variant of Astrophytum asterias. Body solitary, flat globular in shape, 5 to 10 cm tall, 6 to 10 cm wide. 8 broad ribs, green in color with yellow spots. Areoles arrange vertically on the edge of the ribs. Flowers funnel~shaped, pale yellow, 3 to 4 cm in diameter, appear in summer.

黄兜锦
Astrophytum asterias 'Variegata'

鸾凤兜
Astrophytum asterias × A. Myriostigma

为兜和鸾凤玉的杂交种。植株桶形,刺座长有绒毛但无刺。具6条肥厚的棱。
The hybrid of Astrophytam asturias and A. Myriostigma. Body cylindrical, areoles spineless but covered with tomentum. 6 fleshy and thick ribs.

瑞凤玉
Astrophytum capricorne

植株单生，桶形，常见有 8 条较深的薄棱。刺座长于棱缘上，具 6~10 枚长而金黄色的周刺，2 枚金黄色或褐色的中刺。花顶生，橙黄色，花心红棕色。原产墨西哥。

Solitary. Body cylindrical. 8 thin ribs, with areoles growing on the edges. 6 to 10 long and golden radial spines, 2 golden or brown central spines. Flowers terminal, orange, with reddish brown center. Native to Mexico.

五棱碧鸾凤玉
Astrophytum myriostigma huolum

植株扁球形，呈五角星状，碧绿色。刺座长于棱背上。只有一小团绒毛但无刺。花金黄色，花心红棕色，长于植株顶部。是一栽培变种。

A cultivar. Body flat~globular in shape and emerald in color. Areoles grow on the rib edges and are covered with tomentum but not spines. Flowers grow near the apex, and are golden on the outside but reddish brown at the center.

四棱鸾凤玉
Astrophytum myriostigma var. quadricostatum

植株桶形，具有四条肥厚棱，棱缘上长有细小绒毛刺座，无刺，为鸾凤玉四棱变种。

A variant of A. asterias. Body cylindrical. 4 fleshy and thick ribs with tiny, fluffy areoles growing on the edges. Spineless.

鸾凤玉（僧帽）
Astrophytum myriostigma Lem.

花为黄色，花期为 5~8 月，分布于墨西哥中部到南部的高地。

Flowers yellow, produced from May through August. Widespread but scattered in the nrothern and central highlands of Mexico.

四角碧鸾凤玉
Astrophytum myriostigma var. quadricostatum 'Nuclum'

植株桶形，具四条肥厚碧绿色的棱，全株没有灰白色点状绒毛，棱缘无刺。花顶生，黄色。非常受栽种者喜爱。

Body cylindrical. 4 fleshy, thick, and aquamarine ribs. The edges of the ribs are spineless. Flowers terminal and yellow. Very popular species.

六棱鸾凤玉
Astrophytum myriostigma var. 'Hexagenus'

植株桶形，具 6 条肥厚棱，棱缘长有细小绒毛刺座，无刺。为鸾凤玉变种。原产墨西哥。

A variant of A. asterias. Body cylindrical. 6 fleshy and thick ribs with tiny, fluffy areoles growing on the edges. Spineless.. Native to Mexico.

碧琉璃鸾凤玉
Astrophytum myriostigma var. 'Nudum'

植株球形至圆筒形，五条肥厚碧绿色棱，棱缘长有较密绒刺座。无刺。为鸾凤玉的一个变种。花顶生，漏斗形，黄色。相当雅致，为栽培者喜爱。

A variant of A. asterias.Body globular or cylindrical. 5 fleshy, thick, and aquamarine ribs with fluffy areoles growing on the edges. Spineless. Flowers terminal, yellow and funnel-shaped.

七棱鸾凤阁
Astrophytum mytiostigma var. columuare 'Heptagonus'

植株柱形，灰色，七条锐棱。无刺。花顶生，黄色。鸾凤玉的一个变种。

A variant of Astrophytum mytiostigma. Body columnar and grey, 7 sharp ribs. Spineless. Flowers terminal, yellow.

白云般若
Astrophytum ornatum (DC.) Web.

植株倒卵形，具8条棱，刺黄色，刺座长于棱背上，整个植株长有白色绒毛。由于生长环境和栽培条件不同，绒毛疏密不同。花金黄色，花心橙色，顶生为般若变种。

Body obovate, 8 ribs. Spines are yellow in color, areoles grow on the rib edges. White tomentum cover whole body. Flowers are golden in color with orange center.

群凤玉
Astrophytum senile Fric.

植株桶形，长有 8~10 条棱。刺座长有 8~10 条向内弯的周刺，1~3 条褐色至黑色中刺。花顶生，橙黄色，花心红色。原产墨西哥。

Body cylindrical. 8 to 10 ribs. 8 to 10 radial spines, which bend inwards, grow on the areoles. 1 to 3 brown or black central spines. Flowers terminal and orange with red center. Native to Mexico.

白云般若
Astrophytum ornatum var. bubescente

植株单生，球形至圆筒形。株高 20~25 cm，株幅 10~15 cm。具 8 条深尖棱，每条棱生有 3~4 个刺座。表面青灰色。周刺 6 枚，中刺 1 枚，淡黄色。花顶生，漏斗形，黄色。

Solitary. Body globular or cylindrical. 20 to 25 cm high, 10 to 15 cm wide. 8 deep and sharp ribs with 3 to 4 areoles growing on them. Surface bluish grey. 6 radial spines, 1 yellow central spine. Flowers terminal, yellow, and funnel~shaped.

金色火炬（天鹅绒仙人柱、碧彩柱、丝绒柱）（碧彩柱属）

Bergerocactus emoryi

分布在加利福尼亚州南部，以及墨西哥北部的加利福尼亚半岛。

伟冠柱
Borzicactus roezlii (Haage) Backeb.

植株灌木状，茎粗 7 cm，灰绿色。7~14 条棱。周刺 9~14 枚，浅褐色。中刺一根，长 1~4 cm。花靠顶侧开，红色，花蕊黄色，高出花瓣上。原产南美洲秘鲁，现已被广泛引种栽培。

Shrub like. Stem diameter 7 cm, greyish green. 7 to 14 ribs. 9 to 14 radial spines, pale brown. 1 central spine, 1 to 4 cm long. Flowers red and lateral, grow near the top. Stamens and pistils yellow, extend beyond the pedals. Native to Peru, widely cultivated all around the world.

雨树花冠柱
Borzicactus samnensis F. Ritter

植株灌木状，高可达 1.5 m，茎草绿色，茎粗 5~7 cm。具 6~8 条羽毛状棱，周刺 5~10 枚，长 0.3~1.2 cm。中刺 1~4 枚，长 1.5~2.5 cm。花成排侧开，红色，放射状。原产南美洲秘鲁。

Shrub like. Up to 1.5 m high, stems green, 5 to 7 cm in diameter. 6 to 8 feather~like ribs. 5 to 10 radial spines, 0.3 to 1.2 cm long. 1 to 4 central spines, 1.5 to 2.5 cm long. Flowers red and lateral, appear radially in a row. Native to Peru.

雨树花冠柱
Borzicactus samnensis

植株柱形，高约 1.5 m，草绿色，茎粗 5~7 cm。具 10~12 条羽毛状的棱，周刺 5~10 枚，中刺只有一枚，褐色或黑色。花侧生，管状紫红色。原产秘鲁。

Body columnar, up to 1.5 m tall, green. Stem diameter 5 to 7 cm. 10 to 12 feather~like ribs. 5 to 10 radial spines, 1 brown or black central spine. Flowers lateral, tube~shaped, and purplish red. Native to Peru.

丝毛花冠柱
Borzicactus sericatus

植株柱形，基部分枝后成丛生。直立或半匍匐，具 8~10 条纵向粗直棱。丝毛状的刺布满植株，靠顶密生。刺座密生于棱缘上。8~10 枚细针状刺。原产秘鲁，玻利维亚和厄瓜多尔。

Body columnar. Branched at the base. Erect or semi~creeping. 8 to 10 thick straight ribs which are vertically arranged. Body is covered with wool~like spines. Areoles appear on the rib edges. 8 to 10 tiny needle~like spines. Native to Peru, Bolivia, and Ecuador.

佛塔
Brouningia hertlingiana (Backeb.) buxb.

直立生长，树形仙人掌，最高可达8米，茎粗30 cm。棱常多于18条，刺周围有突起。幼株周刺4枚，周刺随生长可增加到多达30枚。3枚长达8 cm的中刺，具黑色刺尖。花白色，原产南美洲秘鲁。

Tree like and erect. Up to 8 m high, stem diameter up to 30 cm. More than 18 ribs, tubercles grow around the spines. 4 radial spines when young, up to 30 after mature. 3 central spines, up to 8 cm long, tips black. Flowers white. Native to Peru.

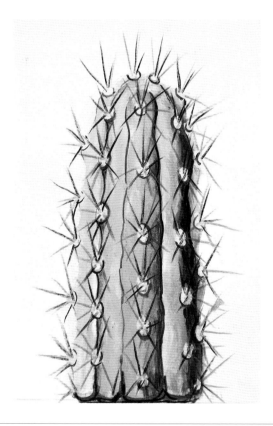

未鉴定仙人掌科植物
Cacti sp.

钢盔天轮柱
Cereus chalybaeus Ottoex C.F. Foist.

植株很少分枝，柱形，具蓝斑。5~6 条深纵向直棱。周刺 7~9 根，中刺 3~4 根。花侧生，瓣外红色，瓣内白色，非常容易开花。原产阿根廷北部和乌拉圭。

Rarely branched. Body columnar, covered with blue spots. 5 to 6 deep straight ribs which are vertically arranged. 7 to 9 radial spines, 3 to 4 central spines. Flowers lateral, outer petal red, inner petal white. Native to northern Argentina, and Uruguay.

姬白云狮子锦
Cereus jamacaru 'Variegata'

又名牙买加天轮柱锦。为牙买加天轮柱的斑锦品种。植株易群生，茎分枝树状，株高 50~100 cm，株幅 15~30 cm，有 4~5 条棱，蓝绿色，镶嵌黄色或全株黄色。刺座着生白色绒毛和黄刺。花漏斗状，白色。花期夏季。

A variant of Cereus jamacaru. Body usually clumping, stems branching and tree like, 50 to 100 cm tall, 15 to 30 cm wide. 4 to 5 ribs, blueish green in color with yellow spots or totally yellow. Areoles are covered with white tomentum and yellow spines. Flower funnel~shaped, white in color, appear in summer.

鬼面阁
Cereus peruvianus

植株高大，乔木状，绿色。每条棱背长满密密麻麻的刺座。周刺6~8枚，黑色，较短。原产南美洲阿根廷，玻利维亚和秘鲁。

Body tall and tree like, green in color. Areoles grow densely on the rib edges. 6 to 8 black, short radial spines. Native to Argentina, Bolivia, and Peru.

山吹
Chamaecereus silvestrii var. aurea

也用 Echinopsis chamaecereus var. aurea 或 Lobivia silvestrii var. aurea，又叫黄体白檀。为白檀的斑锦变种。植株丛生，多分枝，细圆筒形。株高 10~15 cm，株幅 8~10 cm，茎具 6~9 条棱，全体黄色，刺座密生周刺 10~15 枚，白色，无中刺。花侧生，漏斗状，红色，长 7 cm。花期夏季。

A variant of Chamaecereus silvestrii. Body clumping, multi-branched, slim cylindrical in shape, 10 to 15 cm tall, 8 to 10 cm wide, yellow in color, 6 to 9 ribs. 10~15 white radial spines grow densely on the areoles, central spine is absent. Flowers lateral, funnel~shaped, red in color, 7 cm long, appear in summer.

鲜丽玉锦
Chamaelobivia Senrei-Gyoku 'Variegata'

又名红山吹。为白檀与辉凤杂交种的斑锦品种。植株丛生，多分枝，细圆筒形。株高 7~8 cm，株幅 4~5 cm，茎有 10~12 条棱，通体橘红色。刺座着生红褐色细刺。花侧生，漏斗状，红色，花径 2 cm。花期夏季。

Body clumping, multi-branched, slim cylindrical in shape, 7~8 cm tall, 4~5 cm wide, orange~red in color, 10 to 12 ribs. Reddish brown spines grow on the areoles. Flowers lateral, funnel~shaped, red in color, 2 cm in diameter, appear in summer.

阿根廷银毛柱
Cleistoactus smarageliflora (F.A.C. Webber) Br. et R.

植株绿色，具 12~16 条棱，花红色。原产阿根廷和玻利维亚。

Body green. 12 to 16 ribs. Flowers red. Native to Argentina and Bolivia.

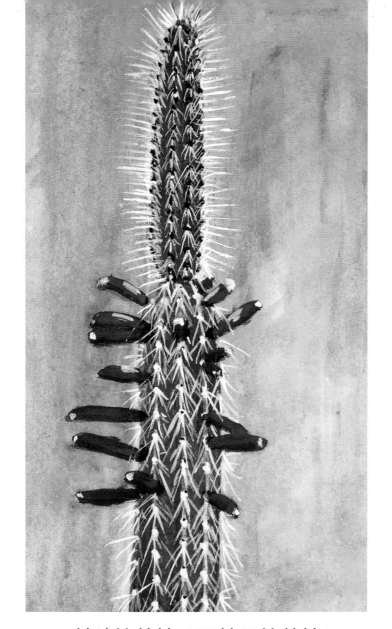

烛焰管花柱（里特里管花柱）
Cleistocactus ritteri Backeb.

植株柱形，翠绿色，具14条或更多的纵向直棱。高达50~60 cm，花侧开，管状。橙红色或红色。原产玻利维亚。

Body columnar, emerald, 50 to 60 cm tall. 14 or more vertical straight ribs. Flowers lateral, tube~shaped, orange or red. Native to Bolivia.

黑士冠
Copiapoa cinered var. De albale Br. et R.

植株呈球形，由不规则疣突组成所谓的棱。单生，暗绿色或墨绿色。刺座生于疣突顶部。具3~5根向两侧伸展的黑色周刺。1~2根黑色较粗壮中刺。花顶生，淡黄色。原产智利。

Body globular, dark green, solitary. The ribs are made of irregular tubercles. Areoles appear on the tip of these tubercles. 3~5 black radial spines which extend to both sides. 1 to 2 thick, black central spines. Flowers pale yellow, terminal. Native to Chile.

百花龙爪玉
Copiapoa grandifiore Ritt.

地中生龙爪玉
Copiapoa hypogaca F. Ritter

植株球形或长球形，绿褐色。表面生有粗大的疣突。株幅可达 6.5 cm，疣突间具棱。茎顶生有白色绵状毛，刺稀少，褐色，后会脱落。花顶生，黄色。原产智利。

Body globular or long globular, greenish brown, up to 6.5 cm wide. Surfaces are covered with big tubercles, and ribs are formed between these tubercles. Spines are brown and sparse. Flowers yellow, and terminal. Native to Chile.

毛疣仙人球
Copiapoa lauii Diers

头部直径小于 2 cm。原产智利 Esmeralda。
Heads are less than 2 cm in diameter. Native Esmeralda, Chile.

鱼鳞玉
Copiapoa tenuissima F.Ritter

植株球形或长球形，接近黑色，宽 5 cm，球体具疣突。刺座生于疣突顶部，并生有小量白色棉状绒毛。周刺 8~16 根，灰白色。原产智利。

Body globular or long globular, nearly black, 5 cm wide. Surfaces are covered with tubercles. Areoles locate on the tip of the tubercles and are covered with white tomentum. 8~16 greyish white radial spines. Native to Chile.

鱼鳞丸
Copiapoa tenuissima F. Ritt.

原产智利。
Native to Chile.

黑王玉
Copinapoa cinerea var. columna-alba

植株初生为球形，成长后逐渐长成圆筒形，灰绿色。具14~30条纵向浅直棱。刺座黑色，披有白色粉状绒毛。生有1~7根褐色针状刺。花顶开，黄色。原产智利。

Body globular when young, gradually turn into cylindrical when mature, greyish green. 14 to 30 vertical shallow ribs. Areoles black, and covered with white tomentum. 1 to 7 brown, needle-like spines. Flowers yellow, terminal. Native to Chile.

金黄恐龙角
Corryocactus anrens (Meyen.) Hutehison

植株直立生长，圆柱状，栽培种茎能长达 50 cm，茎粗 3~5 cm。约 5~8 条棱。周刺 9~11 根，中刺 1 根，长可达 6 cm，刺褐色或黑色。花侧生，橘黄色或红色。原产秘鲁。

Erect. Body columnar. The stems of cultivated are up to 50 cm tall, stem diameter 3 to 5 cm. 5 to 8 ribs. 9 to 11 radial spines, 1 central spine, up to 6 cm long, brown or black. Flowers lateral, orange or red. Native to Peru.

菠萝球（天司球）
Coryphantha bumamma (Ehrenb.) Br. et R.

植株球形，青色至深绿色，株幅 10 cm，植株生有大块的疣突。腋间长有白色绵状绒毛，无中刺。花黄色，花心红色。原产墨西哥。

Body globular with big tubercles growing on the surface, cyan or dark green, 10 cm wide. White tomentum grow under the axil of the tubercles. No central spine. Flowers yellow with red center. Native to Mexico.

喜钙顶花球
Corphantha calipensis H. Bravo

分布于墨西哥。

Distributed in Mexico.

狮子奋迅
Coryphantha cornifera (DC.) Lem.

海胆仙人掌
Coryphantha echinus

分布在美国得克萨斯州西部，球形，株幅 15 cm，高 5 cm。
Distributed in west Texas, body globular, 15 cm wide, 5 cm tall.

白豹（纤细顶花球）
Coryphantha gracilis Bremer & Lau

多数单生，茎高达 8 cm，宽 3.5~4 cm。12~18 根周刺，无中刺。花直径 4~5 cm，淡黄色至奶油色。原产墨西哥奇华华州。

Usually solitary, stems are 8 cm high and 3.5 to 4 cm wide. 12 to 18 radial spines, no central spines. Flowers are 4 to 5 cm in diameter, pale yellow to creamy in color. Native to Chihuahua, Mexico.

劳氏顶花球
Coryphantha lauii Bremer

植株单生，球形至长球形，5 cm 宽。18~20 根周刺，9~13 mm 长。1 根中刺，16 mm 长。花直径 4.5 cm，亮黄色。原产墨西哥科阿韦拉州。

Solitary, globular or elongated globular in shape, 5 cm wide. 18 to 20 radial spines, 9 to 13 mm long. 1 central spine, 16 mm long. Flowers are 4.5 cm in diameter, light yellow in color. Native two Coahuila, Mexico.

枪骑士（长须顶花球）
Coryphantha longicornis Boed.

花黄色。原产墨西哥。
Flowers are yellow. Native to Mexico.

圣克鲁斯蜂窝仙人球（顶花球属）
Coryphantha recurvata

分布在阿利桑那州南部，以及墨西哥附近高地。球形，高 10-25 cm，株幅 7~15 cm，高 3.8 cm。

粗刺顶花球
Coryphantha robust

花为金黄色到浅黄色，分布于美国亚利桑那州南部、新墨西哥州、德克萨斯州和墨西哥索诺拉州、奇华华州。

The colors of flowers range from golden yellow to pale yellow. Native to southern Arizona, New Mexico, Texas, the U.S., and Sonora and Chihuahua, Mexico.

菠萝（球）仙人掌（顶花球属）
Coryphantha robustispina

分布在阿利桑那州南部，新墨西哥州，得克萨斯州，以及墨西哥靠近美国附近。

萨迪顶花球
Coryphantha salm-dyckiana (Scheer) B. & R.

原产墨西哥奇华华州。

Native to Chihuahua, Mexico.

具槽顶花球
Coryphantha sulcata (Englm.) B. & R.

植株丛生，子球 12 cm 宽。周刺 12~14 根。初生没有中刺，成熟后具一根中刺，向外弯曲生长。花直径 5 cm，黄色具红色花心。原产德克萨斯州南部。

Clumping, heads to 12 cm wide. 12 to 14 radial spines. No central spine when young, has one central spine after mature, curved outwards Flowers 5 cm in diameter, yellow in color with red center. Native to south Texas.

花精丸
Coryphantha sulcolanata (Lem.) Lem.

原产墨西哥。
Native to Mexico.

隐柱天轮柱
Cylindropuntia prolifera

隐柱天轮柱（分叉）
Cylindropuntia versicdor

火焰球（变种）
Denmoza erythrocephala (K. Schnm) A. Berger.

植株球形，20~30条直棱，长有30或更多狐红色针刺。原产南美洲阿根廷。

Body globular. 20 to 30 straight ribs. 30 or more red, needle~like spines. Native to Argentina.

火焰球
Denmoza erythrocephala (Salm-Dyck.) Br. et R.

植株长球形，高达 1.5 米，株幅 15~30 cm。具 20~30 个棱角，长有 30 根或更多的刺。刺呈狐红色，长达 6 cm。开花时长出无数白色鬃状核绵状绒毛。花靠顶开，红色。原产阿根廷。

Body long globular. Up to 1.5 m tall, 15 to 30 cm wide. 20 to 30 ribs, about 30 or more red spines, up to 6 cm long. Red flowers appear near the tip. Native to Argentina.

茜球
Denmoza rhodacantha (Slam-Dyck) Br. et R.

植株球形，高与株幅均为 9~16 cm，长有 15 条锋利的纵向直棱，棱间有深沟。周刺 8~10 根，中刺 1 根或没有，黄红色或棕红色，长约 3 cm。花靠顶生，红色，原产阿根廷。

Body globular, 9 to 16 tall and wide. 15 sharp, vertical straight ribs with deep grooves between each of them. 8 to 10 radial spines, 0 to 1 yellowish red or reddish brown central spine, about 3 cm long. Red flowers appear near the tip. Native to Argentina.

赫氏圆盘玉
Discocactus horstii Buin&Bred.

七刺圆盘玉
Discocactus heptacathus (Rodr.) Bret R.

植株高 7 cm，株幅 15 cm。扁平球形，10~19 条肥厚直棱。刺座有白色绒毛，无刺。花白色顶开。原产玻利维亚。

Up to 7 cm tall, 15 cm wide. Body flat globular. 10 to 19 fleshy straight ribs. Areoles spineless and covered with white tomentum. Flowers white and terminal. Native to Bolivia.

碟形圆盘玉
Discocactus insignis Pfeiff.

花灰白色，原产巴西。

Flowers are off~white. Native to Brazil.

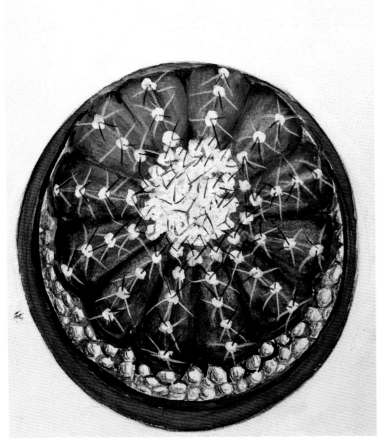

尼逊双重仙人掌
Disocacta nelsonii

植株属于附生类叶花仙人掌，花浅红色，花管较长，花蕊白色突出于花瓣之外。原产中美洲洪都拉斯和危地马拉。

Epiphytic cacti. Flowers pale red, floral tube relatively long, stamens and pistils extend beyond the pedals. Native to Honduras and Guatemala.

金鯱
Echinocactus grusonii Hildm

植株球形或扁球形，具18~20条直棱，周刺25枚。在棱缘刺座上相互交错生长。长约3 cm。花多靠基部生长，大红。果实橘红或大红。原产南美洲阿根廷。

Body globular or flat globular. 18 to 20 straight ribs with areoles growing on the edges. 25 radial spines, up to 3 cm long. Flowers mostly grow at the base, red. Fruits orange or red. Native to Argentina.

红花双重仙人掌
Disocacta phyllanthoides. Br. et R.

为附生类叶花仙人掌，花红色生长于叶两侧。原产中美洲洪都拉斯和危地马拉。

Epiphytic cacti. Flowers red and grow on both sides of the leaves. Native to Honduras and Guatemala.

裸鯱
Echinocactus grusonii 'Breris-niruis'

植株球形，单生，为金鯱短刺栽培种，是一大型种类，具28~32条纵向直棱。刺短密生于棱缘上。花极少开放。

A cultivar of E. grusonii with short spines. Body globular. Solitary. 28 to 32 vertical straight ribs. Short spines thickly grow on the rib edges. Flowers rarely appear.

短刺金鯱
Echinocactus grusonii 'Tansikinshuehi'

又称王金鯱，为金鯱一个栽培变种。植株球形，株高8~10 cm，株幅10~15 cm。茎18~22条棱，但排列不规则。刺座长有白色绒毛，周刺短而粗壮。花顶开，钟状，黄色。

Body globular, 8 to 10 cm tall, 10 to 15 cm wide. 18 to 22 irregular ribs. White tomentum grow on the areoles, radial spines short and thick. Flowers terminal, bell~shaped, yellow.

花王丸
Echinocactus horizonthalonius Lem

植株黄绿色，具8~10条扁平棱。刺座长于棱背上。周刺6~8枚，无中刺。红棕色，较粗壮。向内弯曲生长。原产美国南部和墨西哥北部。

Body yellowish green. 8 to 10 flat ribs. Areoles grow on the rib edges. 6 to 8 thick, brownish red radial spines which bend inwards. No central spine. Native to southern U.S. and northern Mexico.

巨金鯱
Echinocactus ingens

植株球形，单生。株高1.25 m，茎粗1.5 m。20~24条疣节较深的锐棱。刺座长于棱节上，周刺4~5枚，灰白至灰褐色刚刺，1~2枚长2~2.5 cm中刺。花顶生，红色。

Body globular, solitary. Up to 1.25 m high, stem diameter 1,5 m. 20 to 24 deep sharp ribs. 4 to 5 radial spines, off~white or greyish brown, bristle~like. 1 to 2 central spines, 2 to 2.5 cm long. Flowers terminal, red.

岩（广利球）
Echinocactus platyacatus Link et Otto.

植株单生，高 1.5 m，株幅达 1.25 m，为一大型种类。长球形，灰青绿色。具有 50 多条纵向直棱。刺座密生，周刺短 8 根，中刺一根，所有刺均向内弯曲。花小金黄色。原产墨西哥，现已被广泛引种栽培。

Solitary. Up to 1.5 m high, 1.25 m wide. Long globular, greyish cyan. 50 vertical straight ribs. 8 radial spines, 1 central spine, every spines bend inwards. Flowers small and golden. Native to Mexico, and widely cultivated all around the world.

狂刺凌波
Echinocactus texensis 'Hybride'

植株深黄绿色，扁球形。具 10~12 条扁平棱，刺座长于棱背上。长有 5~6 枚红棕色较粗壮弯刺。1~2 枚中刺，为一较受栽种者喜爱的栽培种。

Body dark yellowish green, flat-globular in shape. 10 to 12 flat ribs, areoles grow on the edges. 5 to 6 thick, brownish red curved spines. 1 to 2 central spines.

粗刺凌波
Echinocactus texensis 'Hybride'

植株翠绿，球形。具 10~12 条较锐棱。长有 6~10 枚棕红色粗刺。植株顶部长有多条长 6~8 cm 红棕色中刺。原产墨西哥中北部，深受栽种者喜爱。

Body globular and emerald in color. 10 to 12 relatively sharp ribs. 6 to 10 thick spines which are brownish red in color. Several reddish brown central spines, 6 to 8 cm in length, grow on the apex. Native to the north central Mexico.

凌波
Echinocactus texensis Hopff.

植株高达 15 cm，30 cm 宽。花直径 6 cm。原产美国德克萨斯州和新墨西哥州。

Body up to 15 cm high and 30 cm wide. Flowers 6 cm in diameter. Native to Texas and New Mexico.

Echinocereus adustus Englm.

中刺 1 根，长，黑色。原产墨西哥奇华华州。

1 central spine, long and black in color. Native to Chihuahua, Mexico.

紫红玉
Echinocereus adustus var. schwarzii (Lau) N.P. Taylor

1~5 根中刺。花通常比原种要大。原产墨西哥杜兰戈州。

1 to 5 central spines. Flowers are usually larger than those of the original species. Native to Durango, Mexico.

巴氏鹿角柱
Echinocereus barthelowanus B. & R.

植株丛生，8~10 条棱，花紫色，直径 5~7 cm。原产墨西哥。
Clumping, 8 to 10 ribs, Flowers purple, 5 to 7 cm in diameter. Native to Mexico.

金龙
Echinocereus berlandieri (Eng.) Hort. F.A. Haage

茎匍匐，1.5~5 cm 粗。原产美国德克萨斯州南部以及墨西哥。
Stems creeping, 1.5~5 cm in thickness. Native to south Texas, U.S., and Mexico.

Echinocereus bristolii var. pseudopectinatus N.P. Taylor

植株单生，12~15 根周刺。原产墨西哥。
Solitary, 12 to 15 radial spines. Native to Mexico.

赤寿山鹿角柱
Echinocereus chisoensis W.T. Marsh

植株单生，花直径 5 cm，粉洋红色，具白色喉部。变种 var. fobeanus (Oehme) N.P. Taylor (illus.) 丛生，花 7~12 cm 宽。原产美国德克萨斯州。

Solitary. Flowers 5 cm wide, pinkish magenta in color, with white throat. var. fobeanus (Oehme) N.P. Taylor (illus.), clumping, flowers 7 to 12 cm in diameter. Native to Texas.

御旗
Echinocereus dasyacanthus

花为柠檬黄色,也有橙色。分布于美国亚利桑那州的尤马和比马镇。

Flowers lemon yellow, orange, or pink in color. Nativeto Yuma, and Pima county, Arizona.

司虾(幻虾、武勇丸)
Echinocereus engelmannii (Parry ex Englm.) Lem.

原产墨西哥西北部以及美国西南部。

Native to northwestern Mexico and southwestern U.S.

尼克鹿角柱
Echinocereus engelmannii var. nicholii Benson

原产墨西哥。

Native to Mexico.

九刺虾
Echinocereus enneacanthus Englm.

茎15 cm宽，中刺8 cm长。变种var.brevispinus茎更细，中刺更短，只有5 cm长。原产墨西哥北部。

Stems up to 15 cm wide, central spines are 8 cm long. var.brevispinus has slimmer stems, central spines are also shorter, 5 cm in length. Native to northern Mexico.

九刺虾杂交变种
Echinocereus enneacanthus-Hybrida

植株长球形至筒形，具 8~10 条浅直棱。8~12 枚针状周刺和 1~2 枚中刺。花喇叭形，鲜红色，较大。为一杂交栽培种，颇受栽种者喜爱。

Body shapes range from elongated~globular to cylindrical. 8 to 10 shallow, straight ribs. 8 to 12 needle~like radial spines and 1 to 2 central spines. Flowers are trumpet~shaped, bright red, and relatively large. A hybrid cultivar.

卫美玉
Echinocereus fendleri (Engelm.) Rumpler

富丽鹿角柱
Echinocereus fendleri (Englm.) Ruempler.

植株翠绿色至绿色，长球形，10~12条棱。刺座长在棱背上，中刺1~2枚，周刺6~8枚。能在基部长出小球。花大，鲜红色，侧生。原产墨西哥北部，现已被广泛引种栽培。

Body emerald or green, elongated~globular in shape. 10 to 12 ribs with areoles growing on the edges. 1 to 2 central spines, 6 to 8 radial spines. Able to split at the base. Flowers big, bright red, lateral. Native to northern Mexico, and widely cultivated all around the world nowadays.

幻虾
Echinocereus ferreirianus H. Gates

植株丛生，茎高40 cm，宽8 cm。通常有4根中刺。花具有淡绿色或白色柱头。原产墨西哥。

Clumping, stems up to 40 cm high and 8 cm wide. Usually 4 central spines. Flowers has pale green or white stigmas. Native to Mexico.

幻虾（变种）
Echinocereus ferreirianus H.E. Gates var. lindsayi (Meyran) N.P. Taylor

11~14 条棱，周刺 8~14 根，中刺 4~7 根，长达 10 cm。花紫丁香色，喉部红色，长 6~10 cm。原产墨西哥。

11 to 14 ribs, 8 to 14 radial spines, 4 to 7 central spines, up to 10 cm long. Flowers are purplish violet with red throat, 6 to 10 cm long. Native to Mexico.

明矾
Echinocereus grandis B. & R.

花白色，淡黄色，或者淡粉色。原产加利福尼亚海湾群岛。

Flowers are white, pale yellow, or pale pink. Native to the islands of the Gulf of California.

大花鹿角柱杂交变种
Echinocereus Hybrida

植株筒形，碧绿色，8~10条浅直棱。花靠顶侧生，花特大，粉红色。为栽种者十分喜爱的品种。为一杂交变种。

A hybrid. Body cylindrical in shape, and emerald in color. 8 to 10 shallow, straight ribs. Flowers huge, lateral and appear near the apex, pink in color.

钩刺鯱
Echinocereus johnsoni

宇宙殿
Echinocereus knippelianus Liebner

植株圆柱形，绿色。6~8 条浅棱。刺座杂布，基本看不见刺。花大侧开，大红。原产墨西哥。

Body green and columnar. 6 to 8 shallow ribs. No obvious spines. Flowers red and big, lateal. Native to Mexico.

劳氏鹿角柱
Echinocereus lauii Frank

原产墨西哥。

Native to Mexico.

海胆（魔剑丸、豪剑丸）
Echinocereus leucanthus N.P. Taylor

茎 3~6 mm 粗。花白色，直径 4 cm。原产墨西哥。
Stems 3 to 6 mm thick. Flowers are white, 4 cm in diameter. Native to Mexico.

王将虾
Echinocereus longisetus (Englm.) Lem.

花紫粉色具白色喉部。原产墨西哥科阿韦拉州。
Flowers are purplish pink with white throat. 6 cm wide. Native to Coahuila, Mexico.

墨西哥微刺虾（变种）
Echinocereus ochoterenae J.G.Ortega

植株球形，黄绿色，具6~8条深棱，周刺不多且较短，花大黄色，长于棱背上。原产墨西哥中北部，现已被广泛引种。

Body globular and yellowish green in color. 6 to 8 deep ribs. Not too many radials spines, and they are relatively short. Flowers big and yellow, growing on the rib edges. Native to north central Mexico, and widely cultivated all around the world nowadays.

春高楼
Echinocereus palmeri B. & R.

具有非常有特色的大直根。原产墨西哥奇华华州。

Develops a characteristic large taproot. Native to Chihuahua, Mexico.

柏玛尼逊鹿角柱
Echinocereus pamanesiorum

茎可达 35 cm 高，8 cm 宽，很少分枝。9~12 根周刺，1 cm 长，黄色至白色。0~2 根中刺，1.7 cm 长，棕色具深色尖部。花粉色。原产墨西哥。

Stems up to 35 cm high and 8 cm wide, sparingly branched. 9 to 12 radial spines, 1 cm long, yellow to white in color. 0 to 2 central spines, 1.7 cm long, brown in color, with darker tips. Flowers pink. Native to Mexico.

Echinocereus papillosus A. Linke ex Ruempler

原产美国德克萨斯州 Lowland 地区。

Native to the Lowland areas of Texas.

三光球
Echinocereus pectinatus (Scheidm.) Englm.

茎可达 35 cm 高，8 cm 宽，很少分枝。9~12 根周刺，1 cm 长，黄色至白色。0~2 根中刺，1.7 cm 长，棕色具深色尖部。花粉色。原产墨西哥。

Stems up to 35 cm high and 8 cm wide, sparingly branched. 9 to 12 radial spines, 1 cm long, yellow to white in color. 0 to 2 central spines, 1.7 cm long, brown in color, with darker tips. Flowers pink. Native to Mexico.

黄太阳
Echinocereus pectinatus var.

植株球形，16~18 条浅直棱。棱背上长有较密刺座。周刺 12~16 枚黄白色针状硬刺。为一栽培变种。

A cultivar. Body globular. 16 to 18 shallow, straight ribs with areoles growing densely on the edges. 12 to 16 needle-like, hard spines which are yellowish white in color.

刺虾
Echinocereus polyacanthus var. densus (Regel) N.P. Taylor

花特大，直径可达 8 cm，长 14 cm。原产墨西哥西部。

Flowers particularly large, up to 8 cm in diameter and 14 cm in length. Native to western Mexico.

明石丸
Echinocereus pulchellus (Mart.) Sch.

茎直径在 2.5~5 cm 之间。刺通常较短，约 4 mm 长。原产墨西哥。

Stems 2.5~5 cm in diameter. Spines are usually short, around 4 mm in length. Native to Mexico.

姬路丸
Echinocereus pulchellus var. amoenus (Dietr.) Sch.

茎 3~7.5 cm 宽，刺 6 mm 长。原产墨西哥。
Stems 3 to 7.5 cm wide, spines 6 mm long. Native to Mexico.

平和虾
Echinocereus reichenbachii var. baileyi

通常单身，高达 15 cm，宽 5 cm。约有 14 根周刺，白色至棕色，长达 1.2 cm。1~3 根中刺，3 mm 长。花淡紫色，直径不小于 6 cm。原产美国。

Usually solitary, up to 15 cm high and 5 cm wide. About 14 radial spines, white to brown in color, up to 1.2 cm long. 1~3 central spines, 3 mm long. Flowers light purple, 6 cm or more in diameter. Native to the U.S.

花杯
Echinocereus reichenbachii var. baileyi (Rose) N.P. Taylor

14 根周刺，12 mm 长，中刺 1~3 根，3 mm 长，刺棕色至白色。原产美国俄克拉荷马州南部及德克萨斯州。

Radial spines up to 14 in number, 12 mm long. 1 to 3 central spines, 3 mm long. Spines are brown to white in color. Native to south Oklahoma and Texas.

植株长球形，花侧生，花色多样，粉红，花心绿色，花蕊深绿色，为锦照变种。原产美国西南部和墨西哥。

Body long globular. Flowers lateral, various color, center green, stamens and pistils dark green. Native to southwestern U.S. and Mexico.

锦照虾（折墨虾）
Echinocereus reichenbachii var. fitchii (B. & R.) Benson

周刺多达 22 根，7.5 mm 长。1~7 根中刺，长达 9 mm。其独特的具深红色喉部的粉红色花非常吸引人。原产美国德克萨斯州南部以及墨西哥。

Up to 22 radial spines, 7.5 mm in length. 1~7 central spines, up to 9 mm long. The spectacular pink flowers with their crimson throat are particularly attractive. Native to south Texas, U.S., and Mexico.

珠毛柱
Echinocereus schmollii (Weingart.) N.P. Taylor

原产墨西哥。

Native to Mexico.

鲜红虾（富丽虾）
Echinocereus sciurus var. floresii (Bkbg.) N.P. Taylor

有明显中刺，花开在顶端之下。
Obvious central spines, flowers appear below the apex.

大佛殿
Echinocereus sp.

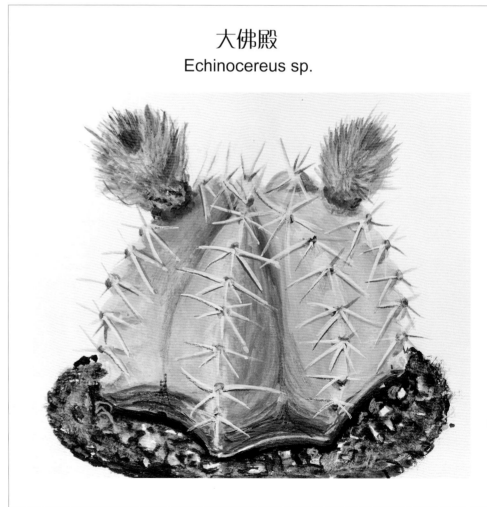

阿巴尼斯虾
Echinocereus spinigemmatus Lau

植株丛生，茎高达 30 cm。原产墨西哥。

Clumping, stems up to 30 cm high. Native to Mexio.

月光虾
Echinocereus stoloniferus subsp. tayopensis

茎长达 15 cm，宽 7.5 cm，通过地下的匍匐枝成丛生长。10~13 根周刺，1.5 cm 长，1~5 根中刺，2.5 cm 长。花黄色，直径 6~7.5 cm。原产墨西哥。

Stems up to 15 cm long and 7.5 cm wide, forming clumps via underground stolons. 10 to 13 radial spines, 1.5 cm long, 1~5 central spines, 2.5 cm long. Flowers yellow, 6~7.5 cm in diameter. Native to Mexico.

荒武者（草虾、集合虾）
Echinocereus stramineus

分布在美国新墨西哥州南部，得克萨斯州西部，以及靠近美国墨西哥附近。筒形，株高 10~30 cm 英寸，株幅 3~10 m。

奥克特大佛殿
Echinocereus subinermis var. ochoterenae (Ortega) Unger

中刺明显，1~4 根。花直径 6.5 cm。原产墨西哥。

Central spines are obvious, 1 to 4 in number. Flowers 6.5 cm in diameter. Native to Mexico.

三刺虾杂交变种
Echinocereus triglochidiatus

青花虾杂交种
Echinocereus triglochidiatus var. melanacanthus-Hybrida

植株球形至柱形，深绿色。14~16 枚带红色针状刺，花顶生，黄色，较大，花心红棕色。为一杂交变种。

Body shapes range from globular to columnar, dark green in shape. 14 to 16 red, needle-like spines. Flowers terminal and relatively big, yellow on the outside and reddish brown at the center.

莫加维皇杯（变种）
Echinocereus triglochidiatus var. mojavensis (Englm. & Big.) Benson

丛生。6~10 根刺，7 cm 长。1~2 根弯曲的中刺。变种 var. gonacanthus (Englm. & Big.) Boissevar. & Davidson 具 6~9 根刺。原产美国莫哈韦沙漠地区以及墨西哥背部。

Forms clumps. 6 to 10 spines, up to 7 cm long. 1~2 central spines, all curving and twisted. var. gonacanthus (Englm. & Big.) Boissevar. & Davidson, 6 to 9 spines. Native to the Mojave Desert area of the U.S., and northern Mexico.

鲜红鹿角柱
Echinocereus triglochidiatus var. neomexicanus (Standley) Benson

植株丛生，11~22 根刺，中刺 4 cm 长。花猩红色、肉粉色，或者白色。变种 var.arizonicus (Rose ex Orcutt) Benson 有 8~10 根刺，中刺 1~2 根，2 cm 长。分布于美国亚利桑那州东南部，新墨西哥州南部和西部，德克萨斯州西部，以及墨西哥

Clumping, 11 to 22 spines, central spine 4 cm long. Flowers are scarlet, salon~pink, or white. var.arizonicus (Rose ex Orcutt) Benson has 8 to 10 spines including 1 to 2 central spines, 2 cm long. Distributed in southeast Arizona, south and west New Mexico, west Texas, and Mexico.

美桃虾
Echinocereus viereckii Werd

茎黄色，6~9 根疣突棱。刺很细。花紫罗兰粉色，直径 11 cm。原产墨西哥。

Stems are yellow, 6 to 9 tuberculate ribs. Spines are very thin. Flowers violet-pink, 11 cm in diameter. Native to Mexico.

青花虾
Echinocereus viridiflorus Englm.

刺红色，棕色，浅灰色，或者白色。周刺 4.5 mm 长。变种 var.correllii Benson 具有黄绿色或灰白色刺，周刺长达 9 mm。花具有柠檬香味。原产美国南达科他州，怀俄明州东南部，以及德克萨斯州。

Spines are red, brown, pale grey, or white. Radial spines 4.5 mm long. var.correllii Benson has yellowish green or off-white spines, radial spines up to 9 mm long, flowers smell like lemon. Native to South Dakota, southeast Wyoming, and Texas.

戴维青花虾
Echinocereus viridiflorus var. davisii (Houghton) W.T. Marsh

体型很小的一个品种，野生种只有 1.2~2 cm 高，0.9~1.2 cm 宽。栽培种体型较大，丛生。原产美国德克萨斯州。

A tiny species, wild type is only 1.2 to 2 cm high and 0.9 to 1.2 cm wide. Cultivated species are larger in shape and clumping. Native to Texas.

鬼见城
Echinoereus triglochidiatus

植株长形，分枝，绿色，刺不多。花大侧生鲜红色。原产美国南部和墨西哥北部。

Body green and branched. Flowers lateral and bright red in color. Native to southern U.S. and norther Mexico.

瓦诺克美刺球
Echinomastus warnockii Glass & Foster

植株单生，11 cm 高。花直径 2.5 cm。原产美国德克萨斯州，新墨西哥州以及墨西哥。

Solitary, up to 11 cm high. Flowers 2.5 cm in diameter. Native to Texas, New Mexico, U.S., and Mexico.

鹿角柱属杂交种
Echinomastus 'Hybrida'

弯刺仙人球
Echinopsis ancistrophora Speg.

一小型、扁平品种，8 cm 宽。花 12~16 cm 长。原产阿根廷。

A small, flattened species, up to 8 cm wide. Flowers 12 to 16 cm long. Native to Argentina.

芳春丸
Echinopsis ancistrophora

植株扁平，深绿色，宽不小于 8 cm。3~7 根周刺，1~4 根中刺，均较短。花白色，5~16 cm 长。原产阿根廷。

Body flattened, dark green in color, 8 cm or more in width. 3 to 7 radial spines, 1 to 4 central spines, all fairly short. Flowers white, 5~16 cm in length. Native to Argentina.

金城柱
Echinopsis camdicans

植株圆柱形，高约 1 m，多为群生。具直棱 11~13 条。每个刺座长有 10~15 枚黄色周刺，中刺 4 枚，长约 1~1.5 cm，黄色。茎顶长有红色针刺。花白色漏斗状。原产南美洲阿根廷、巴西等。

Body columnar. Up to 1 m tall. Usually clumping. 11 to 13 straight ribs. 10 to 15 yellow radial spines grow on every areoles. 4 yellow central spines, 1 to 1.5 cm long. Red needle~like spines grow on the tip of stems. Flowers white and funnel-shaped. Native to Argentina and Brazil.

卡氏弯刺仙人球
Echinopsis cardenasiana

植株单生 10 cm 宽。花侧生，8~10 cm 长，直径 6~7 cm，花洋红色或猩红色。原产玻利维亚。

Solitary, 10 cm wide. Flowers lateral, 8~10 cm long, 6~7 cm in diameter, magenta or scarlet in color. Native to Bolivia.

世界图（地图宝）
Echinopsis eyriesii 'Variegata'

植株幼时为球，成长后渐变为圆筒形。植株高 10~12 cm，株幅 8~10 cm。具 11~12 条较锐的直棱。球绝大部分为黄色，中间有小部分为绿色。着生在刺座上周刺短而粗刚刺。花侧生，白色漏斗状。傍晚开放。

Body globular when young, become cylindrical after mature. 10 to 12 cm tall, 8 to 10 cm wide. 11 to 12 sharp straight ribs. Most parts of the globe are yellow with a small portion of green. Short, thick, bristle~like radial spines grow on the areoles. Flowers lateral, white and funnel~shaped, appear in the evening.

短刺仙人球
Echinopsis eyriesii (Turp) Zucc.

植株球形，球径 17~25 cm，具 12~16 条纵向深直棱，黄色。刺座生于棱背上，较稀疏，并有微疣。生有 4~8 根短黑色周刺。原产巴西和阿根廷。

Body globular, diameter 17 to 25 cm. 12 to 16 vertical deep ribs, yellow. Areoles appear on the back of the ribs. 4 to 8 short, black radial spines. Native to Brazil and Argentina.

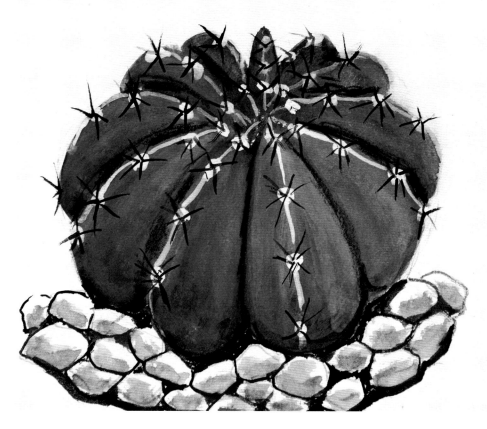

黑刺短毛丸
Echinopsis eyriesii var. (Hybride)

赤盛丸
Echinopsis haematantha (Speg.) D. R. Hunt

绯盛丸
Echinopsis hertrichiana(Backeb.) D. R. Hunt

仙人球属杂交种
Echinopsis-Hybrida (Red)

植株扁球形，绿色。具10~12条浅棱，长有8~12枚周刺，紧贴植株生长。花侧开，红色，较大。原产南美洲中南部。

Body flat-globular, green. 10 to 12 shallow ribs. 8 to 12 radial spines which are tightly attached to the body. Flowers relatively big and red in color, lateral. Native to south central South America.

稀刺仙人球
Echinopsis minuana Speg.

植株黄绿色，长球形，高 80 cm，株幅 15 cm，为一较大型种类，具 12~16 条纵向锐棱，刺座稀小。每条棱背生有 5~6 个刺座。5~6 根黑色周刺。茎顶的刺为灰白色。原产阿根廷。

Body yellowish green, long globular, up to 80 cm tall, 15 cm wide. 12 to 16 vertical sharp ribs, with 5 to 6 areoles growing on the edges. 5 to 6 black radial spines. The spines growing on the stem tip are greyish white. Native to Argentina.

鲜凤丸（红凤丸、鸡冠掌）
Echinopsis mamillosa

植株单生，半球形，宽大于 15 cm。所有刺均为棕黄色，11~16 根周刺，长达 1.2 cm，4~6 根中刺，长达 2.5 cm。花白色或粉色，18 cm 长，直径 9 cm。原产阿根廷。

Solitary, hemispherical. More than 15 cm wide. Spines are all yellowish brown in color, 11 to 16 radial spines, up to 1.2 cm long, 4 to 6 central spines, 2.5 cm in length. Flowers white or pink, up to 18 cm long and 9 cm in diameter. Native to Argentina.

十棱仙人球
Echinopsis molesta Speg.

植株单生，球形，20 cm 宽。花 20~24 cm 长，白色。原产阿根廷 Cordoba。

Solitary, globular in shape, up to 20 cm wide. Flowers 20 to 24 long, white in color. Native to Cordoba, Argentina.

红百合仙人球
Echinopsis obrepanda var. calorubra (Card.) Rausch

丛生，各子球约 14 cm 宽。植株绿色。原产玻利维亚中部。

Forms large, flattened clumps with individual heads up to 14 cm wide. Body green in color. Native to central Bolivia.

白弯刺仙人球
Echinopsis sp.

仙人球属一种
Echinopsis smrziana Backeb.

托氏仙人球
Echinopsis tortispina

植株丛生，具有较长、扭曲的刺。花白色。原产阿根廷 Entre Rios 地区。
Clumping, spines fairly long and twisted. Flowers white. Native to Entre Rios, Argentina.

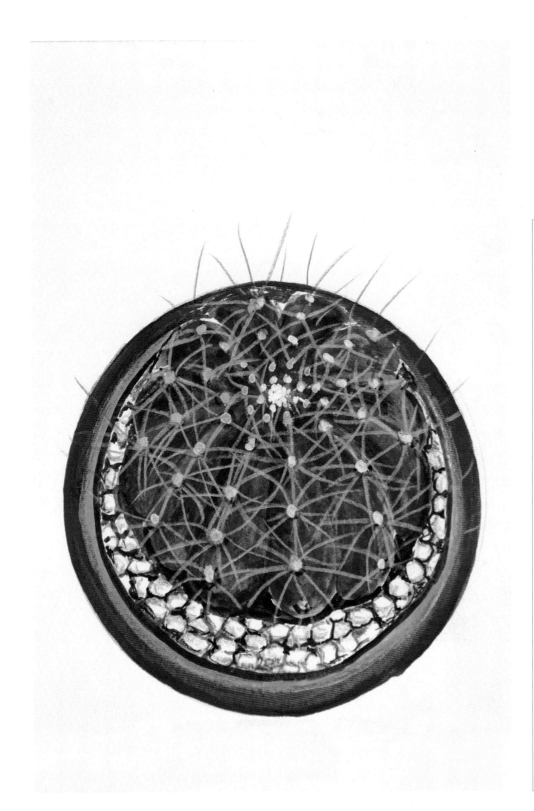

仙人球属杂交种
Echinopsis-Hybrida (Red)

植株扁球形，绿色。具 10~12 条浅棱，长有 8~12 枚周刺，紧贴植株生长。花侧开，红色，较大。原产南美洲中南部。

Body flat~globular, green. 10 to 12 shallow ribs. 8 to 12 radial spines which are tightly attached to the body. Flowers relatively big and red in color, lateral. Native to south central South America.

黄心昙花
Epiphyllum chrysocardium Alexander

植株直径可达 30 cm，分枝。刺座偶尔生有 2~3 根刚毛。花外瓣暗紫粉色，内瓣白色。花长 32 cm。原产墨西哥。

Body up to 30 cm in diameter, branching. Areoles sometimes have 2 to 3 bristle~like spines. Flowers dark purplish pink on the outside, white on the inside, up to 32 cm long. Native to Mexico.

黑闪光（呵云玉）
Eriosyce ausseliana Ritt.

植株扁球形至圆球形，株高 20~25 cm，株幅 20~25 cm。12~20 条不规则疣突棱，灰绿色。周刺 7~9 枚，黑色，针状向内弯曲。花顶生，钟状，洋红色。原产南美洲智利。

Body flat globular or globular, 20 to 25 cm tall and wide. 12 to 20 greyish green ribs are made or irregular tubercles. 7 to 9 black, needle~like radial spines which bend inwards. Flowers bell~shaped and magenta, terminal. Native to Chile.

黑闪光
Eriosyce susseliana

亦称呵云玉。植株扁球形至球形。株高 20~25 cm，株幅 20~25 cm。12~20 疣突棱，灰绿色。周刺 7~9 枚，黑褐色。花顶生，钟状，洋红色。原产南美洲智利。

Body flat globular or globular. 20 to 25 cm tall, 20 to 25 cm wide. 12 to 20 ribs with tubercles, greyish green. 7 to 9 blackish brown radial spines. Flowers terminal, bell~shaped, magenta. Native to Chile.

李氏须弥丸
Escobaria leei Boed

茎较纤细。刺向内弯曲生长。花棕粉色。原产美国新墨西哥州。

Stems relatively slim. Spines bend inwards. Flowers are brownish pink in color. Native to New Mexico.

紫王子
Escobaria minima (Baird) D. R. Hunt

玫瑰红松球
Escobaria roseana (Boed.) Schmoll ev Bux

植株球形，高 5 cm，株幅 3 cm，是个小型种类，周刺 15 根，中刺 4~6 根，黄色。花不大，红色、白色、或咖啡色。原产墨西哥。

Body globular. Up to 5 cm tall, 3 cm wide. 15 radial spines, 4 to 6 central spines, yellow. Flowers are not very big, red, white, or dark brown. Native to Mexico.

横纲
Escobaria vivipara (Nutt) Buxb.

植株多分枝，球形到长球形。周刺 12~40 根，白色或褐色，中刺 1~6 根，褐色。花顶生粉红色。原产美国和加拿大，是一种较耐寒的种类。

Multi-branched. Body globular or long globular. 12 to 40 white or brown radial spines. 1 to 6 brown central spines. Flowers pink and terminal. Native to U.S. and Canada. A cold tolerant species.

秘鲁幻乐
Espostoa melanostele

植株很少分枝，柱形，高2米以上。茎粗4~6 cm，具20~24条较浅的直棱，周刺幼小多达90~120根，白色绒毛状。花靠顶侧生，花白色，花座长有较浓郁白色和褐色的绒毛。原生长在秘鲁北部。

Usually non-branched. Body columnar, more than 2 m tall. Stem diameter 4 to 6 cm. 20 to 24 shallow straight ribs. Radial spines white, tiny, and fluffy, numbering from 90 to 120. Flowers white and lateral, grow near the tip. Thick white and brown tomentum grow on the cephalium. Native to northern Peru.

白丽翁（立翁柱）
Espostoopsis dybowskii

植株高达4 m，基部分枝。周刺短但数量多，隐藏在厚厚的白色绒毛之下。2~3根黄色中刺，长达3 cm。花白色，钟型，4 cm长。原产巴西。

Up to 4 m high, branching at the base. Radial spines short but great in number, concealed beneath the abundant white tomentum. 2 to 3 central spines, yellow in color, up to 3 cm long. Flowers white, bell-shaped, 4 cm long. Native to Brazil.

鹰之巢
Eulychnia acida

植株树形，具 1 m 高树干，植株可达 7 m 高。周刺和中刺混在一起，长达 20 cm。花白色或粉色，直径 5 cm。原产智利。

Tree like with a woody trunk grows up to 1 m high, body is up to 7 m high. Radial and central spines are mixed together, up to 20 cm long. Flowers white of pink, 5 cm in diameter. Native to Chile.

白银城
Eulychnia saint-pieana

植株分枝，灌木状，2~4 m 高。8~12 根周刺，0.5~2 cm 长。1~2 根中刺，5~10 cm 长。花白色，直径 5~7.5 cm。原产智利。

Branching, bush-like, 2 to 4 m high. 8 to 12 radial spines, 0.5 to 2 cm long. 1 to 2 central spines, 5 to 10 cm long. Flowers white, 5 to 7.5 cm in diameter. Native to Chile.

龙王球
Fercocactus echidne (DC.) Br. et R.

植株多单生。球形或桶形，老株则呈柱状。高 35~100 cm （成年植株），株幅 20~30 cm。具 13~21 条纵向直棱，周刺 7~9 根，中刺 1 根，顶生中根红色。花顶生，花色多样，常见黄色。原产墨西哥，现已被广泛引种栽培。

Solitary. Body globular or cylindrical. Columnar after mature, 35 to 100 cm tall, 20 to 30 cm wide. 13 to 21 vertical straight ribs. 7 to 9 radial spines, 1 central spine. Flowers terminal, various colors, usually yellow. Native Mexico, and widely cultivated all around the world.

黄花强刺球
Ferocactus acanthodes

龙虎
Ferocactus echidne

植株柱形，具 18~20 条直棱，具波浪形的棱缘，刺座长于波浪顶端上。周刺 5~6 短，刚刺，刺红色。原产墨西哥东北部和美国西南部。

Body columnar. 18~20 straight ribs with wave~like edges. Areoles grow on the tip of these edges. 5 to 6 short,
red radial spines, bristle~like. Native to northeastern Mexico and southwestern U.S..

狂刺江守
Ferocactus emoryi Cu

植株球形，灰绿色。刺座长于肥厚疣突棱上。6~10 枚粗壮棕黑色直刺。最长超过 10 cm。原产墨西哥。

Body globular and greyish green. Areoles grow on the edges of fleshy tuberculate ribs. 6 to 10 thick, and straight spines which are brownish black in color, and up to 10 cm in length. Native to Mexico.

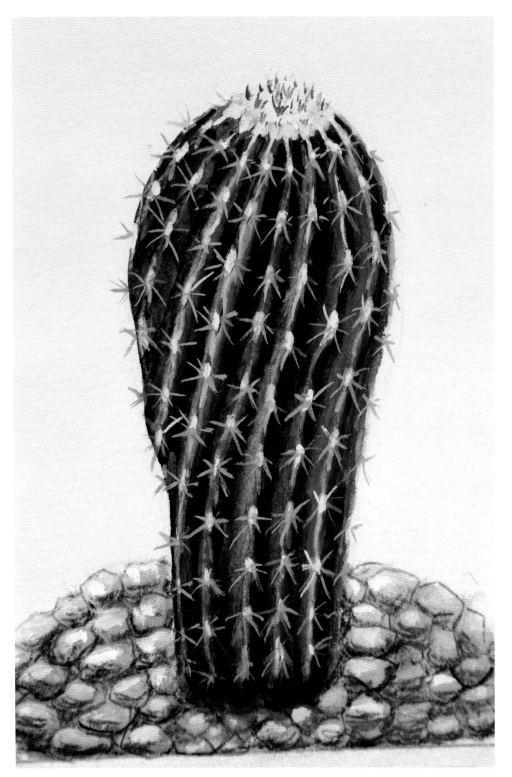

黄绿强刺球
Ferocactus flavovirens (Scheidw.) B. & R.

茎可达 40 cm 长，20 cm 宽。花红色。原产墨西哥。
Stems up to 40 cm high and 20 cm wide. Flowers red in color. Native to Mexico.

福氏强刺球
Ferocactus fordii

直刺强刺球
Ferocactus haematacanthus (S-D) H. Bravo-H

茎 30~120 cm 高，26~36 cm 宽。花紫粉色。原产墨西哥。

Stems 30 to 120 cm high and 26 to 36 cm wide. Flowers are purplish pink. Native to Mexico.

弯刺强刺球
Ferocactus hamatacanthus (Muehlpf.) B. & R.

多数单生，高约 60 cm，宽 30 cm，具圆棱。花黄色，通常具有红色喉部，直径 7.5 cm。原产美国新墨西哥州东南部，德克萨斯州西部和南部，以及墨西哥中北部。

Usually solitary, up to 60 cm high and 30 cm wide, with round ribs. Flowers yellow, usually has red throat, up to 7.5 cm in diameter. Native to southeast New Mexico, west and south Texas, and north central Mexico.

大蜓
Ferocactus hamatacanthus var. hamatidens Br. et R.

植株球形，成长后渐变成长球形或圆筒形。为大虹一个长钩刺的变种，原产美国西南部和墨西哥。也是被广泛引种到各地。花顶生，柠檬黄色。

A variant of F. hamatanthus. Body globular, gradually turn into long globular or cylindrical after mature. Flowers terminal, bright yellow. Native to southwestern U.S. and Mexico. Widely cultivated all around the world nowadays.

大虹
Ferocactus hamatacanthus (Mue Wpf.) Br. et R.

植株球形，成长后变长球形，株高达 60 cm，株幅 30 cm。具 12~17 条疣突棱。周刺 8~12 根，红褐色或灰白色。中刺 4~8 根，刺尖有钩。原产美国西南部和墨西哥，现已广泛被引种到各地。

Body globular, become long globular when mature. Up to 60 cm tall, 30 cm wide. 12 to 17 ribs which are made of tubercles. 8 to 12 reddish brown or greyish white radial spines. 4 to 8 central spines with hooked tips. Native to southwestern U.S. and Mexico. Widely cultivated all around the world nowadays.

文鸟
Ferocactus histrix (DC.) G. Lindsoy

植株球形，植株高达 110 cm，为一大型种类。生有 20~40 条纵向直棱，周刺 6~9 根，中刺 1~4 根，但刺细小，琥珀黄或褐色。原产墨西哥。

Body globular, up to 110 cm tall. 20 to 40 vertical straight ribs. 6 to 9 radial spines, 1 to 4 central spines. Spines are tiny, amber or brown. Native to Mexico.

金鸥玉
Ferocactus latispinus (Haw.) B. & R.

植株单生，高达 30 cm，宽 40 cm。花直径 4 cm，紫色或黄色。变种 var.spiralis (Karw. Ex Pfeiff) N.P. Taylor 具有较小的花，花白色具粉色条纹。广泛分布于墨西哥中部。

Solitary, up to 30 cm high and 40 cm wide. Flowers 4 cm in diameter, purple or yellow in color. var.spiralis (Karw. Ex Pfeiff) N.P. Taylor has smaller flowers, flowers are white with pink stripes. Widely distributed in central Mexico.

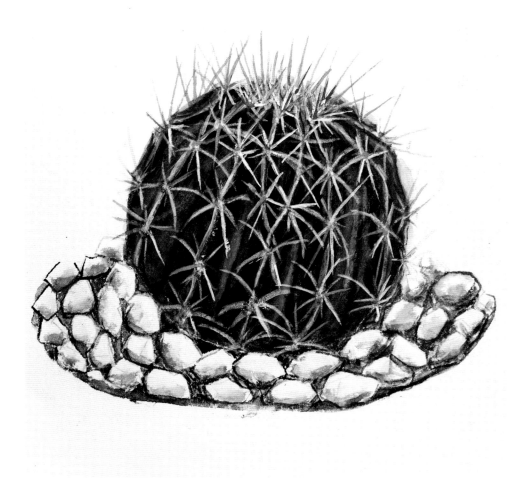

天城
Ferocactus macrodiscas (Mart.) Br. Et R.

植株扁球形，直径最宽可达 40 cm，具 13~25 条纵向直棱，周刺 6~8 根，中刺 4 根，微向内弯。刺黄色或红色，花顶生深红色，紫红色或朱红色。边缘呈现白晕。原产美国。

Body flat globular. Stem diameter up to 40 cm. 13 to 25 vertical straight ribs. 6 to 8 radial spines, 4 central spines which bend slightly inwards. Spines yellow or red. Flowers terminal, dark red, purplish red, or vermilion. Native to the United States.

黄彩玉
Ferocactus schwarzii Lindsay

植株单生，高达 80 cm，宽 50 cm。原产墨西哥。
Solitary, up to 80 cm high, 50 cm wide. Native to Mexico.

毛花筒球
Frailea chiquitema Cardenas.

植株球形，株幅 3~5 cm。由 18~24 条直棱在疣突间散开。周刺 8~12 根，具弯曲，多为深褐色。无中刺或 1~3 根。花黄色顶生。原产玻利维亚。

Body globular, 3 to 5 cm wide. 18 to 24 straight ribs spread between the tubercles. 8 to 12 dark brown, curved radial spines. 0 to 3 central spines. Flowers yellow, terminal. Native to Bolivia.

鳄之子
Frailea horstii Ritt.

植株细长圆柱形，可达 10 cm 高或更多，2.5 cm 宽。花直径 5 cm。原产巴西。

Slender cylindrical, 10 cm or more in height, 2.5 cm wide. Flowers 5 cm in diameter. Native to Brazil.

姬毛玉
Frailea asteriocks Backeb.

植株球形，株幅3 cm。10~15条纵向低平直棱，几乎看不到刺。花顶生，黄色，喇叭形。原产南美洲乌拉圭北部和巴西。

Body globular, 3 cm wide. 10 to 15 vertical straight ribs. No obvious spines. Flowers yellow, trumpet~shaped, terminal. Native to northern Uruguay, and Brazil.

圣王球
Ggmnocactus buenekeri Smales.

植株球形，翠绿色，易在基部长出子球。高6~7 cm，株幅7~10 cm。具5~6条纵向较平，宽阔的直棱。周刺4~5枚，花顶生，钟状，白色至粉红色。原产巴西。

Body globular and emerald. 6 to 7 cm tall, 7 to 10 cm wide. 5 to 6 relatively flat, wide, straight ribs which arrange vertically. 4 to 5 radial spines. Flowers terminal, bell~shaped, colors range from white to pink. Native to Brazil.

绯牡丹锦
Gylmnocalycium mihnovichu var. friedrichii 'Rubra'

植株扁球形，橙红色，5~6条宽阔锐棱。由于植株缺乏光合作用叶绿素，故必须由量天尺等提供养分。为一栽培变种，现已广泛为栽培者喜爱。

Body globular, orange red in color. 5 to 6 wide, sharp ribs. A cultivar.

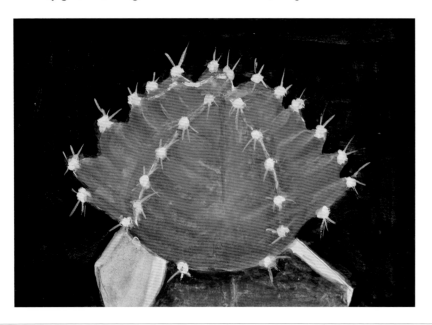

新天地锦
Gymnoalycium saglione 'Variegata'

为新天地的斑锦品种。植株单生，球形，株高10 cm，株幅10 cm。茎具10~13条疣突厚棱，表面深绿色，镶嵌黄色斑块。刺座着生10~12枚周刺，中刺1~3枚，灰褐色至黑红色。花钟型，淡粉红色。花期初夏。

A variant of Gymnoalycium saglione. Body solitary, globular in shape, up to 10 cm tall, 10 cm wide, dark green on the surface with yellow spots. 10 to 13 thick, tuberculate ribs. 10 to 12 radial spines grow on the areoles, 1 to 3 central spines, colors range from greyish brown to blackish red. Flowers bell-shaped, pale pink, appear in early summer.

白花裸萼球
Gymnocalycium stellatum

为新天地的斑锦品种。植株单生，球形，株高 10 cm，株幅 10 cm。茎具 10~13 条疣突厚棱，表面深绿色，镶嵌黄色斑块。刺座着生 10~12 枚周刺，中刺 1~3 枚，灰褐色至黑红色。花钟型，淡粉红色。花期初夏。

A variant of Gymnoalycium saglione. Body solitary, globular in shape, up to 10 cm tall, 10 cm wide, dark green on the surface with yellow spots. 10 to 13 thick, tuberculate ribs. 10 to 12 radial spines grow on the areoles, 1 to 3 central spines, colors range from greyish brown to blackish red. Flowers bell-shaped, pale pink, appear in early summer.

分生裸萼球
Gymnoalycium tillanum

绯花玉（瑞昌玉）
Gymnocactus baldianum (Speg) Speg.

植株球形或扁球形，株高 8 cm，株幅 7~8 cm，茎具 9~11 条纵向浅棱，深绿色，棱中轴有疣突，周刺 5~7 枚，灰白色。花顶生，漏斗状，紫红色。原产阿根廷。

Body globular or flat globular. 8 cm tall, 7 to 8 cm wide. 9 to 11 vertical shallow ribs, dark green. 5 to 7 greyish white radial spines. Flowers reddish violet and terminal, funnel-shaped. Native to Argentina.

黑枪球
Gymnocactus gielsdorfianus

植株球形，株高 5~7 cm，株幅 4~6 cm，茎（球体）具不规则圆锥状疣突构成的所谓的棱，浅绿色至灰绿色。刺座着生细锥形刺 6~8 枚，深褐色。花靠生或顶生，钟状白色或红色。原产墨西哥，为濒危种类。

Body globular, 5 to 7 cm tall, 4 to 6 cm wide. Ribs are made of irregular conical tubercles, color ranges from pale green to greyish green. 6 to 8 thin, conical spines grow on the areoles. Flowers bell~shaped, white or red. Native to Mexico, an endangered species.

Gymnocalycium ferrarii Rausch

植株单生，扁球形。3~4 cm 高，9 cm 宽，灰绿色。花 4.5 cm 长，直径 3.5 cm，白粉色。原产阿根廷 Catamarca.

Solitary, flat~globular in shape. 3~4 cm high, 9 cm wide, greenish grey in color. Flowers 4.5 cm long and 3.5 cm in diameter, whitish pink. Native to Catamarca, Argentina.

宝卵球
Gymnocalycium megalothelos (Seuke) Br. et R.

宽达 16 cm，周刺 7~8 枚，弯曲和紧贴植株生长。花粉红色，顶生。原产南美洲巴拉圭。

Up to 16 cm wide. 7 to 8 curved radial spines, tightly attached to the body. Flowers terminal and pink in color. Native to Paraguay.

Gymnocalycium achirasense Till & Schatzl

植株高达6 cm，宽7cm。多多少少具有一些刺。花漏斗状，直径7 cm，白色泛粉。是 G. horridispinum 和 G. monvillei 的杂交种。原产阿根廷 San Luis。

Body up to 6 cm high, 7 cm wide. More or less covered by spines. Flowers funnel~shaped, 7 cm in diameter, off~white with slightly pink. Hybrid of G. horridispinum and G. monvillei. Native to San Luis, Argentina.

碧岩玉
Gymnocalycium ambatoense

植株扁球形。单生或丛生。头部宽约15 cm，5~10 cm 高。10~12 根刺，2~3 cm 长，通常波浪形，淡棕色。花白色具粉色条纹，直径3~4 cm。原产阿根廷。

Body flat~globular in shape. Solitary or clumping. Heads up to 15 cm wide, 5~10 cm high. 10 to 12 spines, 2 to 3 cm long, usually wavy, pale brown in color. Flowers white with pinkish stripe in the center. 3 to 4 cm in diameter. Native to Argentina.

黄蛇丸
Gymnocalycium andreae (Boed.) Bkbg.

花黄色。原产阿根廷。

Flowers yellow. Native to Argentina.

翠皇冠
Gymnocalycium anisitsii (Sch.) B. & R.

植株单生，长达 10 cm。叶片绿色。花白色，4 cm 长。原产南美洲巴拉圭 Rio Tigatiyami 地区。

Solitary, up to 10 cm long, leaf green. Flowers 4 cm long, white. Native to Rio Tigatiyami, Paraguay.

绯花玉（变种）
Gymnocalycium baldianum var. (Speg) Speg.

植株灰绿色，扁球形。12~14 条疣突的直棱。周刺 6~8 枚，灰白色向内弯曲，紧贴植株生长。花顶生鲜红色。为一栽培变种。

A cultivar. Body greyish green in color and flat~globular in shape. 12 to 14 tuberculate, straight ribs. 6 to 8 radial spines which are tightly attached to the body, greyish white in color and bend inward. Flowers terminal and bright red.

快龙丸
Gymnocalycium bodenderianm var. F. Verieg

植株灰绿色，扁球形。12~14 条疣突的直棱。周刺 6~8 枚，灰白色向内弯曲，紧贴植株生长。花顶生鲜红色。为一栽培变种。

A cultivar. Body greyish green in color and flat~globular in shape. 12 to 14 tuberculate, straight ribs. 6 to 8 radial spines which are tightly attached to the body, greyish white in color and bend inward. Flowers terminal and bright red.

罗星丸
Gymnocalycium bruchii (Speg.) Hoss.

植株具有丛生的、小的、布满刺的子球。花白色至粉红色。var.brigettii Piltz 变种刺较少，花亮粉色。原产阿根廷 Cordoba 地区。

Forming large clumps of small, densely spined heads. The color of flowers ranges from white to dark pink. var.brigettii Piltz has far fewer spines exposing the body beneath and bright pink flowers. Native to Cordoba, Argentina.

布氏裸萼球
Gymnocalycium buenekeri Smaks

植株扁球形，长有 6 条肥微突棱，刺座着生于棱背上，并有白色绒毛。周刺 6~12 根，白色。花顶生，粉红色。原产巴西。

Body flat globular. 6 fleshy ribs. Areoles grow on the back of the ribs, and are covered with white tomentum. 6 to 12 white radial spines. Flowers pink and terminal. Native to Brazil.

粉冠丸
Gymnocalycium capillaense

植株略扁平，丛生。头部宽达 6 cm。5 根细周刺，1.2 cm 长，黄白色。花白色，有时略泛粉，直径 6 cm。原产阿根廷。

Somewhat flattened, clumping. Heads up to 6 cm in diameter. 5 radial spines, thin, 1.2 cm long, yellowish white in color. Flowers white, often mixed with pink, 6 cm wide. Native to Argentina.

桃冠球
Gymnocalycium capillaense (Schick) Backeb

植株长球，浅绿色，8~10 条扁平棱。8~10 枚周刺。花顶生，白色，花心微红。原产南美洲阿根廷。

Body pale green, elongated~globular in shape. 8 to 10 flat ribs. 8 to 10 radial spines. Flowers terminal and white in color, slightly red at the center. Native to Argentina.

光琳玉
Gymnocalycium carminanthum Borth & Koop

刺粗壮，向后弯曲，植株绿色。无中刺。原产阿根廷 Catamarca。

Spines thick, curving backward, body green. Central spines are usually absent. Native to Catamarca, Argentina.

良宽
Gymnocalycium chiquitamum Card.

植株扁球形，6~9 cm 宽，灰绿色，经阳光充分照射后逐渐变红。花 6 cm 长，紫粉色。原产玻利维亚 San Jose 地区。

Body flat globular in shape, 6~9 cm wide, greyish-green, turn into a reddish color under sufficient sunlight. Flowers up to 6 cm long, lilac pink. Native to San Jose, Bolivia.

丽蛇丸
Gymnocalycium damsii (Sch.) B. & R.

植株单生，亮绿色，时常伴有红色。花长 5 cm，直径 5 cm，白色，花心红色。var.tucacocense Bkbg. 变种具有不同的分枝习性，但其他变种如 var.centrispinum, var.rotundulum, and var.torulosum 并没有明显不同。原产南美洲巴拉圭。var.tucacocense 变种原产玻利维亚 San Jose 地区。

Solitary, light, shiny green, often flushed red. Flowers up to 5 cm long and 5 cm in diameter, white with a red center. var.tucacocense Bkbg. differs in its prolific branching habit while Backeberg's other varieties var.centrispinum, var.rotundulum, and var.torulosum differ in no significant way from the species. Native to Paraguay. var.tucacocense Native to San Jose, Bolivia.

红花天王球杂交种
Gymnocalycium denuclatum-Hybrida (Red)

植株球形，8~10 条扁平疣突状棱，蓝绿色。花鲜红，长于植株顶部。原产于南美洲中南部。

Body globular and blueish green in color. 8 to 10 flat, tuberculate ribs. Flowers terminal and bright red. Native to the south central areas of South America.

蛇龙丸
Gymnocalycium denudatum(Link&Otto)Pfeiff.

植株球形，8~10 条扁平疣突状棱，蓝绿色。花鲜红，长于植株顶部。原产于南美洲中南部。

Body globular and blueish green in color. 8 to 10 flat, tuberculate ribs. Flowers terminal and bright red. Native to the south central areas of South America.

天王球杂交种
Gymnocalycium denudatum-Hybrida

植株浅绿色，具扁球形疣突棱。周刺 8~10 枚，花顶生，粉红色。原产南美洲南部阿根廷等地。

Body pale green, with flat-globular, tuberculate ribs. 8 to 10 radial spines. Flowers terminal and pink in color. Native to Argentina and other South America countries.

长刺勇将球
Gymnocalycium eurypleurum Ritt.

植株深绿色，球形。6~8条扁平直棱，刺座长于棱背上。周刺棕褐色粗壮。中刺较周刺粗壮，棕褐色。原产南美洲中南部巴拉圭，阿根廷等地。

Body dark green and globular in shape. 6 to 8 flat, straight ribs with areoles growing on the edges. Radial spines thick and brown in color. Central spines are even thicker than radial spines, also brown in color. Native to Paraguay and Argentina.

海王丸
Gymnocalycium fleischerianum Bkbg.

植株球形或长球形。花4cm长，白色具粉色喉部。原产巴拉圭。

Globular or elongated globular in shape. Flowers up to 4 cm long, white with pink throat. Native to Paraguay.

九纹龙
Gymnocalycium gibbosum

植株深绿色至黑色,通常单生。约60 cm高,12 cm宽。周刺7~10根,长达3.5 cm。0~3根中刺,相似长度。花白色,直径6 cm。原产阿根廷。

Dark green to blackish, usually solitary. About 60 cm high, 12 cm wide. 7 to 10 radial spines, up to 3.5 cm long. 0 to 3 central spines, length as similar as radial spines. Flowers white. 6 cm in diameter. Native to Argentina.

翠皇冠
Gymnocalycium griseopallidum Bkbg.

植株淡灰绿色,高3 cm,宽6.5 cm。花灰白色。原产玻利维亚。

Body pale greyish green, up to 6.5 cm wide and 3 cm high. Flowers off~white. Native to Bolivia.

菲车士裸萼球
Gymnocalycium heischerianum Bkbg.

植株长球,绿色至灰蓝绿色。8~10 条宽阔的直棱。长有 8~12 枚向内弯曲周刺。花顶生,浅粉色。原产南美洲乌拉圭和阿根廷。

Body elongated~globular, colors range from green to greyish green. 8 to 10 wide straight ribs. 8 to 12 radial spines which bend inwards. Flowers terminal and pale pink in color. Native to Uruguay and Argentina.

荷氏裸萼球(圣王球)
Gymnocalycium horstii Buin.

植株长球形,绿色至碧绿色。6~8 条宽阔扁平直棱,刺座不多,周刺也只有 4~6 枚,弯曲紧贴植株。花顶生,粉红色。原产南美洲阿根廷和巴拉圭。

Body elongated~globular in shape and colors range from green to emerald. 6 to 8 wide, flat, straight ribs. Not too many areoles. 4 to 6 radial spines which bend inwards and are tightly attached to the body. Flowers terminal and pink in color. Native to Argentina and Paraguay.

雪冠球
Gymnocalycium hyptiacathum (Lem.) Br. et R.

植株球形，株幅 10 cm，周刺 5~9 根，向内弯曲。中刺 0~1 根。花白色，顶开。原产地阿根廷，但尚有争议。已被广泛引种到世界各地。

Body globular, 10 cm wide. 5 to 9 radial spines which bend inwards. 0 to 1 central spine. Flowers white and terminal. Native to Argentina, and widely cultivated all around the world.

丛生裸萼球
Gymnocalycium leeanum

通常丛生。原产乌拉圭和阿根廷。

Usually clumping. Native to Uruguay and Argentina.

多花牡丹玉锦
Gymnocalycium mihanovichii var. 'Variegata'

植株球形，灰褐色，8~10条锐直棱，花顶生，淡粉红色。为一栽培多花形变种。

Body globular, greyish brown in color. 8 to 10 sharp, straight ribs. Flowers terminal, pale pink in color.

牡丹玉锦
Gymnocalycium mihanovichii var. friedrishii 'Variegata'

植株球形至扁球形，灰褐色。8~10条锐直棱。花顶生，淡粉红色。为牡丹玉栽培变种。

A cultivar of Gymnocalycium mihanovichii. Body shapes range from globular to flat~globular, greyish brown in color. 8 to 10 sharp, straight ribs. Flowers terminal and pale pink in color.

瑞云牡丹锦
Gymnocalycium mihanovichii var. friedrichii 'Aurea'

植株扁球形至球形，金黄色。5~6 条扁平宽阔直棱，多分枝。由于植株本身缺乏光合作用叶绿素，故必须用量天尺为枕本进行嫁接才能正常生长。为一栽培变种，现已广泛引种。

Body shapes range from flat~globular to globular, golden in color. 5 to 6 flat, wide, straight ribs. Multi~branched. A widely cultivated cultivar.

碧严玉
Gymnocalycium nigriareolatum Bkbg

植株球形，单生，15 cm 宽。花白色。原产阿根廷。
Body globular, solitary, up to 15 cm wide. Flowers white. Native to Argentina.

黑罗汉丸（紫冠玉、拉根）
Gymnocalycium ragonesii Castellanos

植株小而扁平，灰棕色。花 4 cm 长，奶油白色，具红色喉部。原产阿根廷 Catamarca。

Body small and flat, greyish brown in color. Flowers up to 4 cm long, creamy white with red throat. Native to Catamarca, Argentina.

武碧玉
Gymnocalycium ritterianum Rausch

植株单生或成小丛生长。花 6.5 cm 长，直径 7.5 cm，白色具光泽，喉部紫粉色。原产阿根廷。

Solitary or forming small groups. Flowers 6.5 cm long and 7.5 cm in diameter, glossy white in color with a pale purplish pink throat. Native to Argentina.

云龙（观音龙、纹美玉、纹绯丽）
Gymnocalycium schatzlianum Till & Schatzl

植株扁球形，9~11 cm 宽。成熟后可高达 15 cm，宽 17 cm。花直径 6~8 cm。原产阿根廷 Cordoba。

Body flat~globular, 9 to 11 cm wide. Up to 15 cm high and 17 cm wide after mature. Flowers 6 to 8 cm in diameter. Native to Cordoba, Argentina.

波光球（华龙丸、琥龙丸、琥龙光、新天龙、龙光丸、龙华丸、爬荒龙、蛇斑龙）
Gymnocalycium schickendantzii (Web.) B. & R.

植株单生，宽达 10 cm，成熟后长成圆柱状。刺的形状和数目多变。花长达 5 cm，白色或粉色。原产阿根廷西北部。

Solitary, up to 10 cm wide, become columnar in shape when mature. Spines are variable in shape and number. Flowers up to 5 cm long, white or pink in color. Native to northwestern Argentina.

须黑玉
Gymnocalycium schroederianum van Osten

植株单生，15 cm 宽，5 cm 高。15~18 条棱。花长达 7 cm，直径 5.5 cm，白色。变种 var.bayense Kiesling 具更小的茎，约 7 cm 宽，花 4~5 cm 长。变种 var. paucicostatum Kiesling 只有 9~11 条具棱角的棱，只有 3 根刺。原产乌拉圭。

Solitary, up to 15 cm wide and 5 cm high. 15 to 18 ribs. Flowers up to 7 cm long and 5.5 cm in diameter, white in color. var.bayense Kiesling has smaller stems, mostly up to 7 cm wide. Flowers 4 to 5 cm long. var.paucicostatum Kiesling has only 9 to 11 ribs, very angular, only 3 spines. Native to Uruguay.

土蜘蛛
Gymnocalycium sp.

植株球形至长球形，刺座着生于疣突顶上，周刺 4~6 枚，向下垂生长，白色。

Body globular or long globular. Areoles grow on the tips of the tubercles. 4 to 6 white radial spines, bend downwards.

豪刺花王丸
Gymnocalycium sp.

植株淡绿色至黄绿色。6~8 条扁平直棱,刺座生长 4~6 条粗壮红褐色弯曲的周刺。刺座长有灰白色绒毛。原产南美洲阿根廷等地。

Body colors range from pale green to yellowish green. 6 to 8 flat, straight ribs. 4 to 6 thick, reddish brown, and curved radial spines grow on the areoles. Native to Argentina and other areas of South America.

青玉丸
Gymnocalycium sp.

植株扁球形,翠绿色。具10~12条扁平肥厚直棱,周刺褐色6~10枚。花大黄色,顶生。原产南美洲乌拉圭,巴拉圭等地。

Body flat-globular in shape and emerald in color. 10 to 12 flat, fleshy, straight ribs. 6 to 10 brown radial spines. Flowers big and yellow, terminal. Native to Uruguay, Paraguay, and other areas of South America.

Gymnocalycium sp.

天平丸
Gymnocalycium spegazzinii B. & R.

植株单生，扁球形。可达 20 cm 高，18 cm 宽。刺长短不一，有时紧贴植株生长。花白色，有时带点粉色。原产阿根廷 Salta, Catamarca, 以及 Tucuman 等地。

Solitary, flat globular in shape, up to 20 cm high and 18 cm wide. Spines are variable in length and are usually tightly attached to the body. Flowers white, sometimes with a pink tinge. Native to Salta, Catamarca, and Tucuman, Argentina.

绫鼓（碧盘玉、翠盘玉）
Gymnocalycium tudae var. izozogsii

体型小于原种，但具有跟多的周刺，中刺一根。原产玻利维亚 Santa Cruz.

Smaller than the original species, more radial spines and one central spine. Native to Santa Cruz, Bolivia.

绫鼓（碧盘玉、翠盘玉）
Gymnocalycium tudae var. pseudomalacocarpus (Bkbg.) Don.

植株扁球形，7 cm 宽，3 cm 高。通常红绿色。3~5 根刺。原产玻利维亚。

Body flat~globular, 7 cm wide and 3 cm high. Usually reddish green. 3 to 5 spines. Native to Bolivia.

内弯刺裸萼球
Gymnocalycium uebelmannianum Rausch

植株单生，扁球形，1 cm 高，7 cm 宽，绿色。花长 3.5 cm，直径 3.5 cm，白色具粉色喉部。原产阿根廷。

Body solitary, flat-globular in shape, 1 cm high and up to 7 cm wide, green in color. Flowers 3.5 cm long and 3.5 cm in diameter, white in color with pink throat. Native to Argentina.

守金摩天龙
Gymnocalycium weissianum Bkbg.

植株宽球形，9 cm 高，14 cm 宽，灰绿色。花偏钟型，棕色或粉白色，具深色喉部。原产阿根廷 Mazan。

Broad-globular in shape, up to 9 cm high and 14 cm wide, greyish green in color. Flowers somewhat bell-shaped, brown or whitish pink in color with dark colored throat. Native to Mazan, Argentina.

恐龙球（钟鬼球）
Gymnocalycium horridispinum G. Frank

植株球形或柱形，株幅 8 cm。周刺 10~12 根，中刺 4 根，刺坚硬，金属灰色，刺尖褐色。花顶生，粉红或带粉红和白色。原产阿根廷。

Body globular or columnar, 8 cm wide. 10 to 12 radial spines, 4 central spines. Spines hard, metal grey, spine tips brown. Flowers terminal, pink, or pink and white. Native to Argentina.

长钩玉
Hamatocactus uncinatus

巨锁龙（月亮柱、新桥）
Harrisia martini

茎倾斜，超过 2 m 高，2~2.5 cm 宽。5~7 根短周刺，1 根中刺，2~3 cm 长。花白色，20 cm 长。原产阿根廷。

Stems leaning. Over 2 m long, 2~2.5 cm wide. 5 to 7 radial spines, short. 1 central spine, 2~3 cm long. Flowers white, 20 cm long. Native to Argentina.

砂王女（伊须萝玉）
Islaya krainzina Ritt. Rupe.[*Eriosyce islayensis* (C. F. Först.) Katt.]

光虹球
Lobivia arachnacantha Buin. & Ritt.

植株 4 cm 宽。花 4 cm 长。var. orrecillasensis (Card.) Bkbg. 变种的刺更密，花红色。var.sulphurea Vasqu. 呈亮绿色，花柠檬黄。原产玻利维亚 Samaipata 地区。

Body 4 cm wide. Flowers 4 cm long. var.orrecillasensis (Card.) Bkbg. has denser spines, flowers red. var.sulphurea Vasqu. bright green, flowers lemon~yellow. Native to Samaipata, Bolivia.

Lobivia arachnacantha var. densiseta Rausch

植株密布棕红色长达 20 mm 的刺。花红色，长 7.5 cm。原产玻利维 Valle Grande 地区。

Body is densely covered by brownish red spines which are up to 20 mm long. Flowers red and 7.5 cm long. Native to Valle Grande, Bolivia.

黄裳丸
Lobivia aurea

植株高大，单生或丛生。头部宽达 12 cm，长达 10 cm。8~16 根周刺，长约 1 cm。4 根中刺，长达 3 cm。花黄色，7~9 cm 长。原产阿根廷。

Body high, solitary or clumping. Heads up to 12 cm wide, 10 cm high. 8~16 radial spines, up to 1 cm long. About 4 central spines, up to 3 cm long. Flowers yellow, 7~9 cm long. Native to Argentina.

豪剑丸
Lobivia aurea var. shaferi (B. & R.) Rausch

植株高达 25 cm, 4 cm 宽。中刺 5 cm 长。来自 San Luis 的变种 var. leucomalla (Wessn.) Rausch 是单生，高达 12 cm, 6 cm 宽，密布白色刚毛状刺。原产阿根廷 Andalgala 地区。

Stems up to 25 cm high, 4 cm wide. Central spines 5 cm long. var. leucomalla (Wessn.) Rausch from San Luis is solitary, up to 12 cm high, 6 cm wide, densely covered with white bristle~like spines. Native to Andalgala, Argentina.

湘阳丸
Lobivia bruchii

有时丛生。头部扁球形, 50 cm 宽。9~14 根周刺, 4 根中刺，均 0.5~2 cm 长。花红色，直径 5 cm。原产阿根廷。

Sometimes clumping. Heads flat~globular in shape, 50 cm wide. 9 to 14 radial spines, 4 central spincs, 0.5~2 cm long. Flowers red, 5 cm in diameter. Native to Argentina.

梦春丸
Lobivia caineana

植株单生，亮绿色。高达20 cm，宽达9 cm。花粉色，直径7 cm。原产玻利维亚。

Solitary, bright green in color. Up to 20 cm high and 9 cm wide. Flowers pink, 7 cm in diameter. Native to Bolivia.

粉花芳春丸
Lobivia cardenasiana

植株相对扁平，10 cm宽。周刺12~14根，中刺2~3根。所有刺都约3 cm长。花洋红色或红色，10 cm长。原产玻利维亚。

Body relatively flat. Solitary. 10 cm wide. 12~14 radial spines, 2~3 central spines. All spines are up to 3 cm long. Flowers are magenta or red in color, 10 cm long. Native to Bolivia.

龟甲丸
Lobivia cinnabarina

植株单生或形成少量子球。15 cm 宽。12~16 根刺，其中中刺 1 根，长达 2 cm。花红色，直径 6~8 cm。原产玻利维亚。

Solitary or forming small clumps. Body up to 15 cm wide. 12 to 16 spines with 1 central spine, up to 2 cm long. Flowers red, 6~8 cm in diameter. Native to Bolivia.

Lobivia famatimensis (Speg.) B. & R.

植株单生，宽达 3.5 cm。花直径 3~5 cm。var.sanjuanensis Rausch 变种高达 20 cm，5 cm 宽，花直径 9 cm。var.jachalensis Rausch 变种刺座上绒毛密集，花相对较小。原产阿根廷。

Solitary, up to 3.5 cm wide with a fleshy tap-root. Flowers 3~5 cm in diameter. var. sanjuanensis Rausch up to 20 cm high and 5 cm wide. Flowers up to 9 cm in diameter. var. jachalensis Rausch areoles more woolly, flowers smaller. Native to Argentina.

红花阳盛球（阳盛丸）
Lobivia famatiminsis (Red)[*Echinopsis famatimensis* (Speg.) Werderm.]

红火炬仙人球
Lobivia grandiflora var. crassicaulis (Bkbg.) Rausch

茎高达 15 cm，宽 11.5 cm。花 6~8 cm 长。原产阿根廷 Catamarca。

Stems 15 cm high and 11.5 cm wide. Flowers 6~8 cm long. Native to Catamarca, Argentina.

太刀姬
Lobivia haematantha

植株单生或少量丛生，头部 5~10 cm 高，6~7 cm 宽。5~6 根周刺，1 cm 长。3~4 根中刺，5 cm 长。花红色具白色喉部，直径 5~7 cm。原产阿根廷。

Solitary or sparingly clumping. Heads 5~10 cm high and 6~7 cm wide. 5 to 6 radial spines, 1 cm long. 3 to 4 central spines, 5 cm long. Flowers red with white throat, 5 to 7 cm in diameter. Native to Argentina.

Lobivia haematantha var. amblayensis (Rausch) Rausch

植株多数单生，2 cm 高，3 cm 宽。花直径 6~10 cm，黄色至橙色。原产南美洲阿根廷 Amblya 地区。

Usually solitary, 2 cm high and 3 cm wide. Flowers 6~10 cm in diameter, yellow to orange. Native to Amblaya, Argentina.

Lobivia haematantha var. hualfinensis (Rausch) Rausch

植株单生，7 cm 宽。中刺长达 8 cm。花直径 5.5 cm，橙红色至红色，具白色喉部。原产阿根廷 Catamarca。

Solitary, 7 cm wide. Central spine up to 8 cm long. Flowers 5.5 cm in diameter, orange~red to red in color, with white throat. Native to Catamarca, Argentina.

Lobivia haematantha var. rebutioides (Bkbg.) Rausch

植株在原产地呈单生，栽培后多数丛生。头部 3~5 cm 宽。花直径 6~8 cm，白色、黄色、橙色、红色、粉色、或紫色。原产南美洲阿根廷。

Solitary in the native habitat, but cultivated species are usually clumping. Heads 3~5 cm wide. Flowers 6~8 cm in diameter, white, yellow, orange, red, pink, or violet. Native to Argentina.

红笠丸（朱丽丸、朱丽球）
Lobivia jajoiana

植株单生，6 cm 高，6 cm 宽。9~11 根周刺，1~2 cm 长。1~3 根中刺，3 cm 长，略呈钩状。花番茄红色，直径 6 cm。原产阿根廷。

Solitary, up to 6 cm high and wide. 9 to 11 radial spines, 1 to 2 cm long. 1 to 3 central spines, 3 cm long, somewhat hooked. Flowers are tomato red in color, 6 cm in diameter. Native to Argentina.

Lobivia kieslingii Rausch

植株单生，球状，宽达 25 cm。花直径 9 cm，橙红色至深红色。原产阿根廷 Tucuman 地区。

Solitary, globular in shape, up to 25 cm wide. Flowers 9 cm in diameter, orange-red to carmine in color. Native to the Sierra de Quilmes, Tucuman, Argentina.

劳氏丽花球
Lobivia rauschii Zecher

植株高达 15 cm，宽 5 cm。花直径 4 cm，红色。原产玻利维亚。

Body up to 15 cm high, 5 cm wide. Flowers 4 cm in diameter, red in color. Native to Bolivia.

丽刺丸
Lobivia rosariana Rausch

植株单生，球形，宽达 10 cm。花 6.5 cm 长，直径 5 cm。原产阿根廷 Sierra Famatina 地区。

Solitary, body globular, up to 10 cm wide. Flowers 6.5 cm long, 5 cm in diameter. Native to Sierra Famatina, Argentina.

凄厉丸
Lobivia saltensis (Speg.) B. & R.

植株单生，约 5 cm 宽。刺较细。花 5 cm 宽。var.pseudocachensis 变种植株较小，丛生，刺较原种更密。原产阿根廷。

Solitary, up to 5 cm in width. Spines thin. Flowers 5 cm wide. var.pseudocachensis (Bkbg.) Rausch body small, clumping, spines are denser than those of the original species. Native to Argentina.

优髯球
Lobivia schieliana Bkbg.

植株多头，头部宽达5 cm，深绿色。刺3 cm长，向内弯曲生长。花直径5 cm，红色或黄色。变种var.quiabayensis (Rausch) Rausch头部6 cm宽，绿色，刺5 cm长，花橙色，红色，或洋红色。原产玻利维亚。

Multi~headed, heads are up to 5 cm wide, dark green in color. Spines up to 3 cm long, bend inwards. Flowers 5 cm in diameter, red or yellow in color. var.quiabayensis (Rausch) Rausch, heads 6 cm wide, body green, spines 5 cm long, flowers orange, red, or carmine. Native to Bolivia.

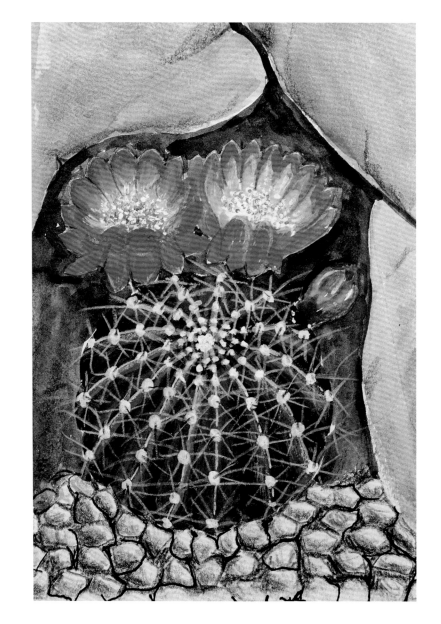

花扇丸
Lobivia schreiteri var. stilowiana (Bkbg.) Rausch

头部7 cm宽。花直径4 cm，橙色或洋红色。变种 var.riolarensis Rausch 植株单生，12 cm宽，花橙色。原产阿根廷。

Heads up to 7 cm wide. Flowers 4 cm in diameter, orange or carmine in color. var. riolarensis Rausch is solitary, up to 12 cm wide, flowers orange in color. Native to Argentina.

花生仙人球
Lobivia silvestrii (Speg.) Rowley

具有多个指状茎。花直径4cm，橙红色至红色。原产阿根廷。

Forms a number of finger-like stems. Flowers 4 cm in diameter, orange-red to red in color. Native to Argentina.

Lobivia sp.

脂粉丸（无忧愁）
Lobivia tiegeliana var.cinnabarina (Fric) Rowley

植株通常单生，宽 7 cm。花直径 3.5 cm，红色带紫。原产阿根廷。

Usually solitary, up to 7 cm wide. Flowers 3.5 cm in diameter, red in color with violet tints. Native to Argentina.

Lobivia wrightiana var. winteriana (Ritt.) Rausch

植株单生，7 cm 粗。花直径 8 cm，亮粉色或橙粉色。原产秘鲁 Villa Azul。

Solitary, 7 cm thick. Flowers 8 cm in diameter, bright pink or orange-pink in color. Native to Villa Azul, Peru,

哲克仙人球
Lobivia zecheri Rausch

通常单生，植株偏圆柱形。0~1 根中刺，8 cm 长。花直径 4 cm。原产秘鲁。
Usually solitary, body somewhat cylindrical in shape. 0 to 1 central spine, up to 8 cm long. Flowers 4 cm in diameter. Native to Peru.

乌羽玉
Lophephora wieeianmsii (Lem. ex Sdm-Dyek) J.M.Coulter

植株单生，蓝绿色，扁球形。株幅 7.5 cm，由扁平疣组成 7~10 条棱，刺座生有黄白色的毡毛团。花粉红色。原产美国西南部和墨西哥。
Solitary. Body blueish green and flat globular. 7.5 cm wide. 7 to 10 ribs which are made of flat tubercles. Areoles are covered with yellowish white wool ball. Flowers pink. Native to southwestern U.S. and Mexico.

翠冠玉
Lophophora cliffusa (Croiz) Bravo.

植株扁球形或球形，株高5 cm，株幅5~7 cm。茎具8个低圆疣突棱，深蓝绿色。刺座着生黄白色绒毛。花顶生，钟状，黄色，故亦称黄花乌羽玉。原产美国西南部和墨西哥。

Body flat globular or globular. Up to 5 cm tall, 5~7 cm wide. Stems have 8 dark buleish green, tubercle-shaped ribs. Areoles are covered with white tomentum. Flowers yellow and bell~shaped, terminal. Native to southwestern U.S. and Mexico.

翠冠玉（吹子）
Lophophora cliffusa (Croiz) Bravo.

原产墨西哥克雷塔罗州。

Native to Queretaro, Mexico.

翠冠玉
Lophophora diffusa (Croiz) Bravo.

也称黄花鸟羽玉。植株圆球形，株高 5 cm，株幅 5~7 cm。8 个低圆疣突棱，蓝绿色。刺座长有黄白色绒毛。花顶生，钟状，黄色。原产美国和墨西哥。

Body globular. Up to 5 cm tall, 5 to 7 cm wide. 8 bluish green ribs with tubercles. Yellowish white tomentum grow on the areoles. Flowers terminal, bell~shaped, yellow. Native to U.S. and Mexico.

地久丸
Malacocarpus erinacaus Lem. Forst

植株单生或分枝。球形或圆柱形，无乳汁。疣突开始有黄色绒毛，周刺 12~20 枚，白色，中刺 0~2 枚。花朱红色，靠顶环开。种子褐色。原产墨西哥。

Solitary or branched. Body globular or columnar. Tubercles are covered with yellow tomentum when young. 12 to 20 white radial spines, 0 to 2 central spines. Flowers vermilion, grow in circle near the top. Seeds brown. Native to Mexico.

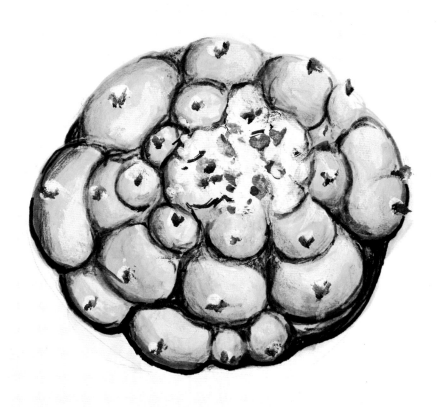

明显乳突球
Maimmillaria conspicua Purpus.

植株球形或长球形，成长后圆柱形，深绿色。具30~40条小锥形疣突排成螺旋状的棱。18~26枚白色短小刚刺包裹球体。2~4枚较长白色中刺。花洋红色，靠顶环形着生。原产墨西哥，现已被广泛引种到各地。

Body globular or long globular, become columnar after mature, dark green. 30 to 40 ribs which are made or spirally arranged, tiny conical tubercles. 18 to 26 white short bristle~like spines cover up the body. 2 to 4 relatively long, white central spines. Flowers magenta, appear in a cycle near the tip. Native to Mexico, widely cultivated all around the world.

白鹭
Mammillaria albiflora (Werd.) Bkbg.

植株成熟后丛生。花白色，直径3 cm。原产墨西哥。

Clumping at the base when mature. Flowers white, 3 cm in diameter. Native to Mexico.

雪头球
Mammillaria chiocephala Purpus.

植株初生为单生，成长后则生出子球，筒形。植株高达20~30 cm。具32~42条排成螺旋状，小锥形疣突的棱，其腋间着生多层白色棉状绒毛环。花小靠顶环形开放，粉红色。原产墨西哥，现已被广泛引种到各地。

Solitary when young, split into several individual when mature. Body cylindrical, 20 to 30 cm tall. 32 to 42 ribs which are made of spirally arranged tiny conical tubercles. White, multi~layer tomentum grow under the axil of the tubercles. Flowers small and pink, appear near the apex. Native to Mexico, widely cultivated all around the world.

樱富士（明日之光）
Mammillaria boolii

通常单生，高达3.5 cm，3 cm宽。约有20条周刺，长1.5 cm，白色。1根钩状黄色中刺，长2 cm。花粉色具灰边，直径3 cm。原产墨西哥。

Usually solitary. 3.5 cm high, 3 cm wide. About 20 radial spines, 1.5 cm long, white in color. 1 central spine, 2 cm long, hooked, yellow in color. Flowers pink with paler margins, 3 cm in diameter. Native to Mexico.

吉赛尔
Mammillaria giselae

茎 1.5~3 cm 宽，形成宽 5~8 cm 的子球。16~21 根周刺，0.5 cm 长。中刺较周刺短，0~5 根。花粉色，直径 1 cm。原产墨西哥。

Stems 1.5~3 cm broad, forming clumps 5~8 cm in width. 16 to 21 radial spines, 0.5 cm long. Central spines are shorter, 0 to 5 in number. Flowers pink, 1 cm in diameter. Native to Mexico.

白豹丸
Mammillaria napina Purpus

多数单生，4~6 cm 宽。花直径 4 cm。原产墨西哥。

Usually solitary, 4 to 6 cm wide. Flowers 4 cm in diameter. Native to Mexico.

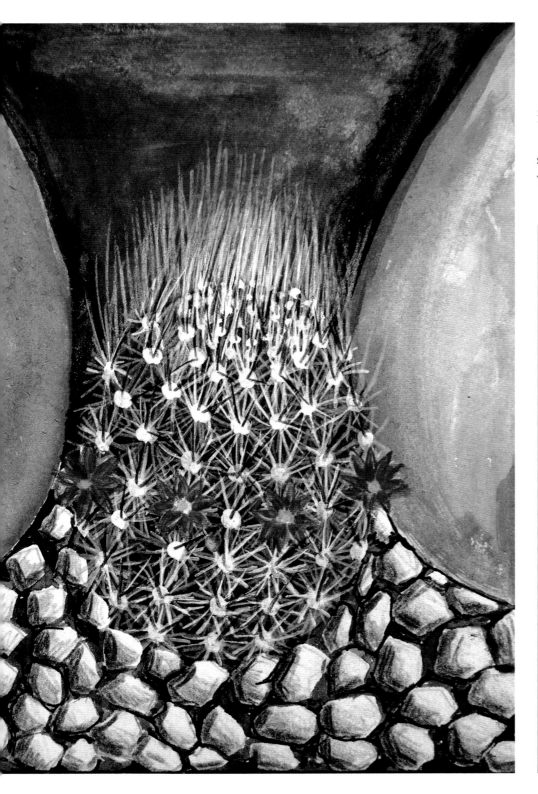

红艳锦
Mammillaria pottsii Scheer ex S-D

植株丛生，茎圆柱形，约 15 cm 高，4 cm 宽。变种 var.multicaulis Repp 的茎较细。变种 var.gigas Repp 较粗，高约 25 cm，宽 7 cm。原产德克萨斯州南部。

Clumping, stems cylindrical, up to 15 cm high and 4 cm wide. var.multicaulis Repp has slimmer stems. var.gigas Repp is thicker, up to 25 cm long and 7 cm wide. Native to south Texas.

猩猩丸
Mammillaria spinosisimna Lem.

植株球形，多单生，无乳汁。刺座具密集的疣突绒毛和刚毛。周刺 20~30 枚，多为白色。也有黄色渐变为浅褐色或红色。花浅朱红色或紫红色，靠顶环开。原产墨西哥。

Body globular, usually solitary. Areoles are thickly covered with tomentum and bristles. 20 to 30 radial spines, mostly white, sometimes yellow or gradually turn into pale brown or red. Flowers vermilion or reddish violet, grow in circle near the top. Native to Mexico.

富贵丸
Mammillaria tetrancistra Englm.

植株丛生，茎高达 25 cm，宽 7.5 cm。分布于美国西南部以及墨西哥西北部。

Clumping, stems up to 25 cm high and 7.5 cm wide. Distributed in southwestern U.S. and northwestern Mexico.

舞衣
Mammillaria wrightii Englm. & Bigelow

植株多数单生，球形至短柱形，宽达 8 cm。花直径 2.5 cm，深粉红色。源于墨西哥奇瓦瓦州 Santa Clara 峡谷的变种 var.wolfii (Hunt) Repp. (illus.) 具有白色花朵。原产美国亚利桑那州和新墨西哥州。

Usually solitary, globular to short-columnar, up to 8 cm wide. Flowers 2.5 cm in diameter, dark pink. var. wolfii (Hunt) Repp. (illus.) from Santa Clara Canyon, Chihuahua, Mexico, has white flowers. Native to Arizona and New, Mexico, U.S.

海仙玉
Matucana crinifera Ritt.

球形或长球形，可达 30 cm 长，10 cm 宽。花 6~7 cm 长。原产秘鲁。

Globular or elongated globular in shape, up to 30 cm high, 10 cm wide. Flowers 6 to 7 cm long. Native to Peru.

白玉仙
Matucana intertexta

植株球形，深绿色，16~20 条浅直棱，长有 8~10 枚周刺。花顶生，橙红色。原产南美洲秘鲁。

Body globular and dark green in color. 16 to 20 shallow straight ribs. 8 to 10 radial spines. Flowers terminal and orange red in color. Native to Peru.

普氏白仙玉
Matucana pujupatii (Don. & Lau) Bregman

植株绿色具疣突棱，刺明显。原产秘鲁。

Body green with tuberculate ribs, spines are obvious. Native to Peru.

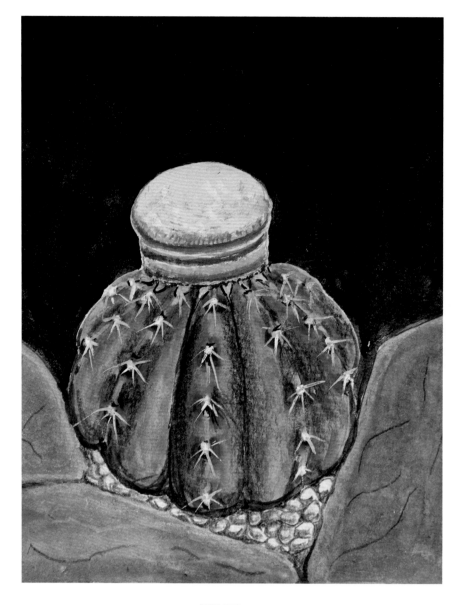

丽云
Melocactus guaricensis

植株柱形，绿色，茎顶花座由褐色灰白色刚毛组成，颜色分层。花座顶部蘑菇形。周刺6~8枚，灰白色。原产加勒比海各岛屿。

Body columnar and green. The cephalium is made of brown and off-white bristles. The tip of the cephalium is mushroom~shaped. 6 to 8 off~white radial spines. Native to the islands of Caribbean sea.

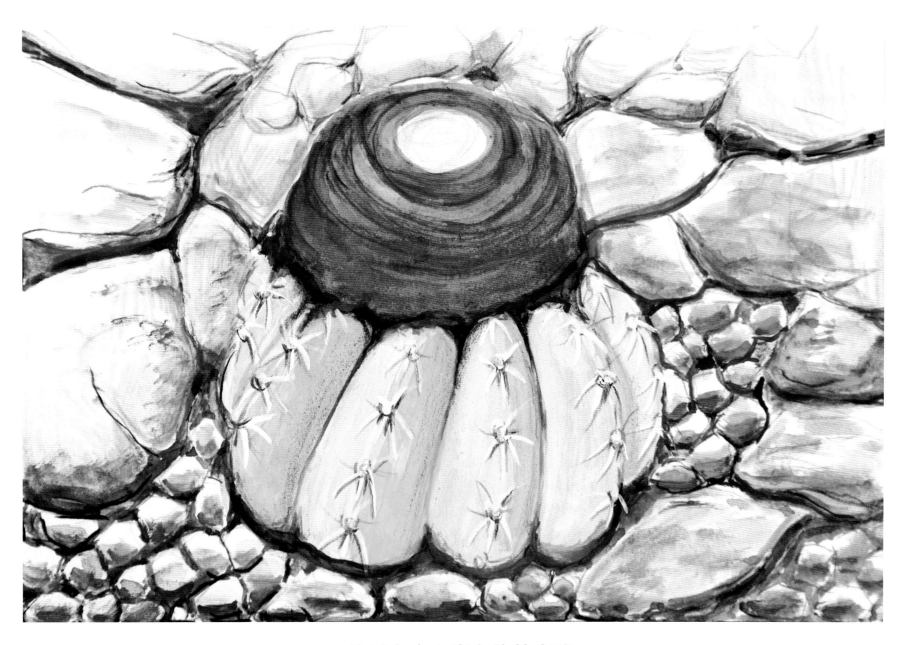

生长在火山岩上的花座球
Melocactus sp.

植株黄绿色，12~14条较肥厚的棱。茎顶生长着塔型的、由棕红色短刚毛组成的花座球。顶部生长白色刚毛。原产加勒比海的岛屿火山岩中。
　　Body yellowish green with 12 to 14 fleshy thick ribs. Tower~shaped cephalium, which are comprised by short, brownish red bristles, grow on the stem tip. White bristles grow on top of the plant. Native to the islands of Caribbean sea, grow on the volcanics.

翠云
Melocactus violacaeus

植株球形，深翠绿色，长有16条直棱，花座圆柱形，长有红棕色短刚毛，颜色深浅分层，十分美丽。周刺6~8枚，灰白色。原产加勒比海的岛上。

Body globular and dark emerald. 16 straight ribs. Short and reddish brown bristles grow on the columnar cephalium. 6 to 8 off~white radial spines. Native to the islands of Caribbean sea.

小花仙人柱
Micranthocereus flaviflorus

底部丛生。茎高75 cm，宽4 cm。植株布满周刺，0.5 cm长。约9根中刺，2 cm长。花橙黄色，红色，或奶油色，管状，1.8 cm长，直径0.6 cm。原产巴西。

Clumping from the base. Stems up to 75 cm high and 4 cm wide. Numerous radial spines, 0.5 cm long. About 9 central spines, 2 cm long. Flowers are yellowish orange, red, or tubular in shape, 1.8 cm long, 0.6 cm in diameter. Native to Brazil.

龙神市
Myrtillocaotus geomertrzaus (Mart.) Console

  树形，生有浓密的冠和短小的根，植株开始为青色，表面如一层薄霜，株高可达 4 m，株幅 6~10 cm，具 5~6 条尖棱，周刺 0~5 根，中刺一根，根褐色或黑色，花白色，果实青紫色，浆果，原产墨西哥，可以用作嫁接枕木。

  Tree like. Body cyan when young, surface looks like to be covered with a thin layer of frost. Up to 4 m high, 6 to 10 cm wide. 5 to 6 sharp ribs, 0 to 5 radial spines, 1 central spine. Roots short, usually brown or black. Flowers white, fruits baccate and bluish violet. Native to Mexico.

勇风
Neobuxbaumia euphorbroide

  植株柱形，8 条直棱，刺座着生棱缘上，覆盖有绒毛装周刺，中刺 1 枚，黑色。花顶侧生，喇叭型，红色。

  Body columnar. 8 straight ribs with areoles growing on the edges. Areoles are covered with fluffy radial spines, one black central spine. Flowers lateral and grow near the top, red in color and trumpet~shaped.

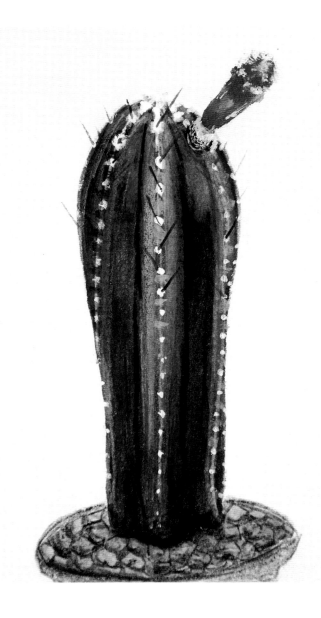

黑冠丸（素妆玉）
Neochilenia paucicostatus Backb.[*Eriosyce paucicostata*(F. Ritter)Ferryman]

大轮丸
Neollogolia conoidea var. (PC.) Br. et R.

植株长球形，高 8~10 cm，株幅 4 cm，很少分枝。具 12~16 条不明显的纵向直棱，刺座有白色绒毛，周刺 16~20 根，白色。中刺 0~1 根。花顶生，深紫红色。原产美国西南部和墨西哥。为都锦的一个变种。

A variant of N. conoidea. Body long globular, 8 to 10 cm high, 4 cm wide, rarely branched. 12 to 16 vertical straight ribs. Areoles are covered with white tomentum. 16 to 20 white radial spines. 0 to 1 central spine. Flowers dark reddish violet, terminal. Native to southwestern U.S. and Mexico.

海氏激光球
Neoporteria heinrichiana (Bkbg.) R.M. Ferryman

植株单生，球形，宽不小于 6 cm。花直径 4 cm，黄色具红尖。原产智利。

Solitary, globular in shape, 6 cm or more in width. Flowers 4 cm in diameter, yellow in color with red tips. Native to Chile.

豹头
Neoporteria napina

植株单生，2~4 cm 宽，4 cm 高。3~9 根周刺，3 mm 长。0~1 根中刺。花淡黄色，3 cm 长。原产智利。

Solitary. 2~4 cm wide, 4 cm high. 3 to 9 radial spines, 3 mm long. 0 to 1 central spine. Flowers pale yellow, 3 cm long. Native to Chile.

雷头玉
Neoporteria occulta (Sch.) B. & R.

植株扁平，宽达 4.5 cm，有时丛生。花 2.5 cm 长，淡黄色。原产智利北部。

Body flattened, up to 4.5 cm wide, sometimes clumping. Flowers 2.5 cm long, pale yellow. Native to northern Chile.

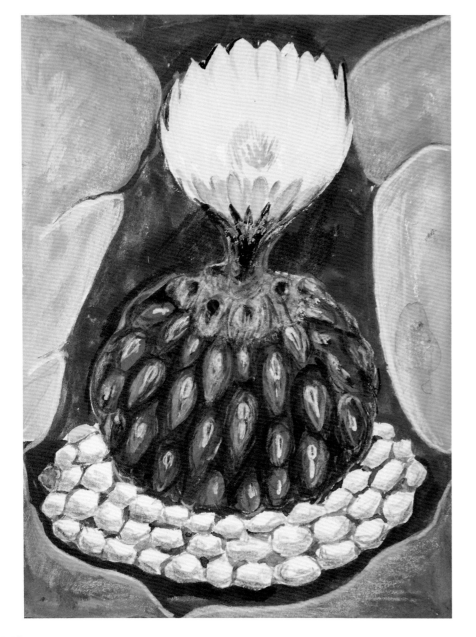

逆龙玉
Neoporteria subgibbosa var. vastanea (Ritt.) R.M. Ferryman

植株单生，球形，逐渐变为长球形。具深棕色或灰白色透明的刺。花 6~7 cm 长。原产智利。

Solitary, globular in shape, gradually become elongated globular. Spines are dark brown or glassy whitish grey. Flowers 6 to 7 cm long. Native to Chile.

短刺智利球
Noeporteria brachyacanthus

植株球形，多单生。具 24~26 条纵向瘤状棱，刺座着生于瘤状顶部。成长植株长有 6~8 根褐色短周刺，花顶生，粉红色，不大。原产于南美洲南部各地。

Usually solitary. Body globular. 24 to 26 vertical ribs. 6 to 8 short, brown radial spines when growing. Flowers pink, terminal, and relatively small. Native to southern part of South America.

令箭荷花
Nopalxochia ackermanii (Haw.) F. M. Knuth.

为附生类仙人掌（也称叶花仙人掌类）生有扁平的叶片状茎，叶多棱。褐绿色，花大鲜红色，原产墨西哥。现被大量引种，并繁育多种花色和栽培种，深受人们喜爱。

Epiphytic cacti. Stems flat and leaf~like, multiple brownish green ribs. Flowers big and bright red. Native to Mexico, and widely cultivated all around the world.

白乐天（白小町）
Notocactus seopa var.

植株倒卵形，易从基部分枝。具 20~24 条纵向浅棱，刺座密生，长有白色绒毛。周刺 10~16 根，细短。茎顶长有白色绒毛，中有红棕色短细刚毛。原产巴西和乌拉圭，为小町个变种。

A variant of Notocacu seopa. Body obovate, usually branched at the base. 20 to 24 vertical shallow ribs. Areoles are covered with white tomentum. 10 to 16 thin and short radial spines. White tomentum, along with reddish brown short bristles appear on the tip of the stems. Native to Brazil and Uruguay.

丛生南国玉
Notocactus caespitosus Speg.

植株丛生，高达 8 cm，宽 3 cm。花 4 cm 长。原产巴西。

Clumping, up to 8 cm high and 3 cm wide. Flowers are 4 cm long. Native to Brazil.

海神丸
Notocactus crassigibbus Ritt.

植株单生，扁平，宽达 10 cm。花约 10 cm 长。原产巴西。

Solitary, body flattened, up to 10 cm in width. Flowers up to 10 cm long. Native to Brazil.

棕色锦绣玉
Notocactus fuscus Ritt.

植株单生，4~7 cm 宽。花不小于 3 cm 长。原产巴西。

Solitary, 4 to 7 cm wide. Flowers 3 cm or more in length. Native to Brazil.

红彩玉
Notocactus herteri Werd.

植株单生，球形或长球形，不小于 15 cm 宽。原产乌拉圭 Cerro Galgo。

Solitary, body globular or elongated globular in shape, 15 cm or more in width. Native to Cerro Galgo, Uruguay.

短刺绒毛南国玉
Notocactus magnificus (Ritt.) Krainz

植株球形，生有 14~16 条纵向直棱，棱缘长有白色绒毛。株高 20 cm，株幅 3~6 cm。花顶生，金黄色。原产巴西。

Body globular. 14 to 16 vertical straight ribs with white tomentum growing on the edges. Up to 20 cm tall, 3 to 6 cm wide. Flowers golden and terminal. Native to Brazil.

英冠玉
Notocactus magnificus (Ritt.) Krainz

植株丛生，宽达 20 cm，球形。花直径 5~6 cm。原产巴西。

Clumping, body up to 20 cm wide, globular in shape. Flowers 5 to 6 cm in diameter. Native to Brazil.

妙梅锦绣玉
Notocactus mueller-melchersii Fric ex Bkbg.

植株单生，高达 8 cm，宽 6 cm。花直径 5 cm，淡金黄色。原产乌拉圭。

Solitary, up to 8 cm high and 6 cm wide. Flowers 5 cm in diameter, pale golden yellow in color. Native to Uruguay.

Notocactus muricatus (Otto) Berg.

植株丛生，子球不小于 20 cm 高。花 3 cm 长。原产巴西南部。
Clumping, heads 20 cm or more in height. Flowers 3 cm long. Native to southern Brazil.

青王球
Notocactus ottonis (Lehm.) Berg.

花直径不小于 6 cm。原产巴西南部，乌拉圭，巴拉圭，以及阿根廷东北部。
Flowers 6 cm or more in diameter. Native to southern Brazil, Uruguay, Paraguay, and northeastern Argentina.

珊瑚城（白毯丸）
Notocactus roseiflorus Schl. & Bred.

植株单生，12 cm 长，6.5 cm 宽。花直径 3.5 cm。原产乌拉圭。

Solitary, up to 12 cm high, 6.5 cm wide. Flowers 3.5 cm in diameter. Native to Uruguay.

红刺南国玉
Notocactus roseolueus var. Vtiet.

植株球形，绿色。具 22~24 疣突锐棱。刺座生在疣突腋下，10~12 枚红色绒毛状周刺。花顶生，粉红色，花蕾朱红色，花蕊黄色。原产南美洲乌拉圭，现各地都有引种栽培。

Body globular and green. 22 to 24 sharp ribs with tubercles. Areoles grow under the axil of the tubercles. 10 to 12 red, fluffy radial spines. Flowers terminal and pink. Flower bud vermilion, stamens and pistils yellow. Native to Uruguay, and widely cultivated around the world.

灰刺南国玉
Notocactus tephraeanthus

植株一般单生，球形或椭圆形，绿色或黄绿色。20~22 条较深的螺旋状具横筋的锐棱。刺座着生于棱缘上。6~8 枚灰白色周刺和一枚灰褐色针状中刺。茎顶生有白色绒毛，花顶生，橙红色。原产南美洲。

Usually solitary. Body globular or oval, green or yellowish green. 20 to 22 relatively deep, spiral~shaped sharp ribs with areoles growing on the edges. 6 to 8 off-white radial spines and 1 greyish brown central spine. White tomentum grow on the stem tip. Flowers terminal and orange. Native to South America.

眩美玉
Notocactus uebelmannianus Buin.

植株单生，高达 12 cm，宽 17 cm。花直径 5 cm，紫色或黄色。原产巴西。

Solitary, up to 12 cm high and 17 cm wide. Flowers are 5 cm in diameter, purple or yellow in color. Native to Brazil.

Notocactus veenianus Van Vliet

植株单生，可达 20 cm 长，8 cm 宽。花直径 5 cm。原产乌拉圭。
Solitary, up to 20 cm high and 8 cm wide. Flowers 5 cm in diameter. Native to Uruguay.

瓦拉斯锦绣玉
Notocactus warasii Ritt.

植株单生，高达 50 cm，宽 20 cm。花直径 6 cm。原产巴西。
Solitary, up to 50 cm high and 20 cm wide. Flowers 6 cm in diameter. Native to Brazil.

黄花海王球（变种）
Notocactus crassigibus

植株翠绿色，8~10 条肥原棱。花较大生长于植株顶部。周刺 8~10 枚，中刺一枚。原产南美洲乌拉圭和巴西。

Body emerald. 8 to 10 ribs. Flowers relatively big and grow on the apex. 8 to 10 radial spines, 1 central spine. Native to Uruguay and Brazil.

密花团扇
Opuntia durangensis

植株长扁平倒卵形，灰绿色至灰褐色。刺座黑色，周刺 4~6 枚，灰白色。花密生于茎背上，金黄色。原产于墨西哥和美国西南部。

Body long, flat, and obovate, color ranges from greyish green to greyish brown. Areoles black, 4 to 6 greyish whit radial spines. Golden flowers appear densely on the back of the stems. Native to southwestern U.S. and Mexico.

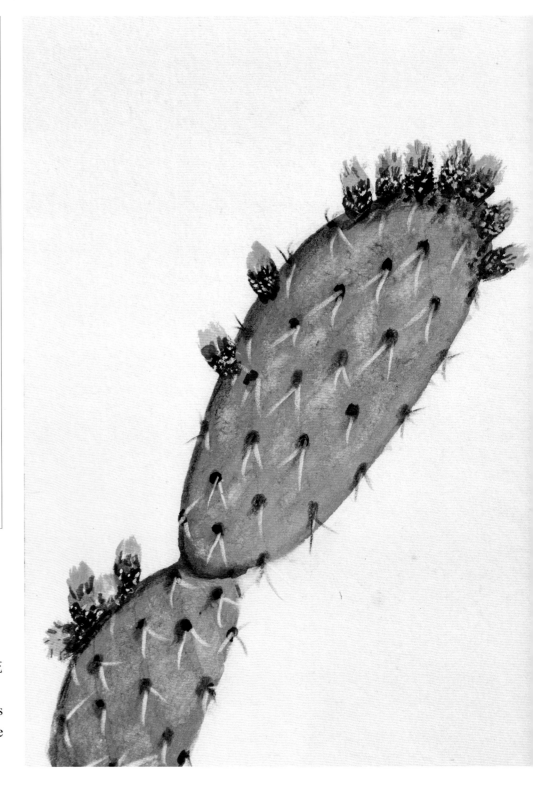

使命仙人掌（金枪无花果、印度无花果）
Opuntia ficus-Indica

黄花仙人掌
Opuntia gosseliniana

白刺仙人掌（白毛掌）
Opuntia microdasys var. albispina Fobe.

植株茎短，偏圆卵形。刺座白色绒毛，无刺。原产墨西哥，现已为各地所引种栽种。

Stems short, and ovate. Areoles are covered with white tomentum, spineless. Native to Mexico, widely cultivated all around the world.

单刺仙人掌
Opuntia monacantha

植株长卵形，扁平，深绿色。刺座只生一枚灰白刺。花橙红色，原产美国西南部和墨西哥北部。

Body long ovate, flat, dark green. Only one off-white spine grow on the areoles. Flowers orange. Native to southwestern U.S. And northern Mexico.

棕色脊椎刺梨（仙人镜）
Opuntia phaeacantha

土人团扇
Opuntia phaecantha Engelm. var. camanchia (Engelm) Borg.

植株扁平，近圆形，翠绿色。花生有团扇缘上。花大，黄色。原产美国西南部和墨西哥。

Body flat globular, emerald. Flowers big and yellow. Native to southwestern U.S. and Mexico.

红刺黄花仙人掌
Opuntia pheacantha Engelmam

植株扁平似手掌，暗黄绿色。周刺5~6根红褐色。花大金黄色。原产美国和墨西哥。

Body dark yellowish green and flat, palm-like. 5 to 6 reddish brown radial spines. Flowers big and golden. Native to U.S. and Mexico.

将军
Opuntia subututa (Muehlenpf.) Engelm.

植株圆柱形，高2~4 m，茎长有突起，粗6~10 cm，肉质粗针形叶12~15 cm长，花生于茎顶部，橙红色，原产阿根廷和玻利维亚。

Body columnar, 2 to 4 m high. Stems diameter 6 to 10 cm, with tubercles growing on the surface. Leaves fleshy and needle-like, 12 to 15 cm long. Flowers grow on the stem tip, orange. Native to Argentina and Bolivia.

清淡仙人掌（银楯）
Oputia azurea

植株扁平，周刺灰白色，6~8根。花较大，金黄色。花心红色。原产美国西南部和墨西哥。

Body flat. 6 to 8 greyish radial spines. Flowers relatively big, golden, with red center. Native to southwestern U.S. and Mexico.

圣云龙
Oreocereus hendriksenianus

丛生。分布于秘鲁高海拔地区。

Clumping. Distributed in high altitude areas of Peru.

武伦柱
Pachycereus pringlei (S. Watson) Br. et R.

株高 12 m，在原产地茎粗可达 1 m，茎生有 10~16 条浅棱，刺约 20 枚。初生红色或褐色，后转为黑色或灰色。长约 2 cm。原产墨西哥。

Up to 12 m high. Stems diameter up to 1 m when growing on native habitats, 10 to 16 shallow ribs. About 20 spines, red or brown in color when young, become black or grey after mature, up to 2 cm long. Native to Mexico.

黄心锦绣玉
Parodia aureicentra Bkbg.

多数丛生，15 cm 宽，刺形状和数量多变。花直径 4 cm，鲜血红色。原产阿根廷北部。

Usually clumping, 15 cm wide, spines are variable in shape and number. Flowers 4 cm in diameter, light blood~red in color. Native to northern Argentina.

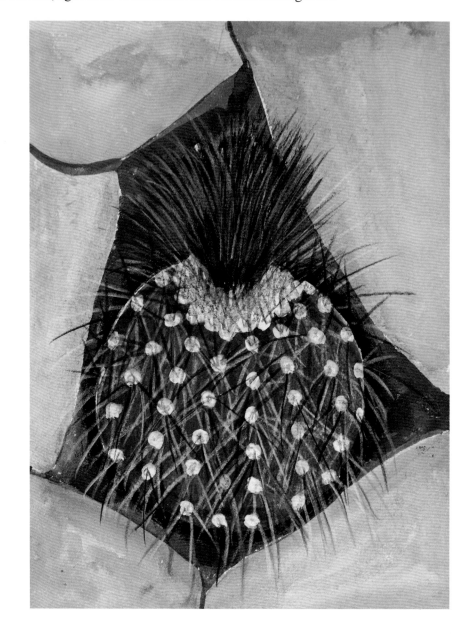

锦绣玉
Parodia aureispina Bkbg.

植株单生，球形，宽达 6.5 cm。花红色。原产阿根廷北部。
Solitary, globular in shape, up to 6.5 cm wide. Flowers red. Native to northern Argentina.

黄刺锦绣玉
Parodia chrysacanthion (Sch.) Bkbg.

植株单生，球形。花小，黄色。原产阿根廷北部。
Solitary, body globular. Flowers are small, yellow in color. Native to northern Argentina.

柱状锦绣玉
Parodia columnaris Card.

植株单生，高达 30 cm，宽 7 cm。花直径 3 cm，亮黄色。原产玻利维亚。

Solitary, up to 30 cm high and 7 cm wide. Flowers are 3 cm in diameter, light yellow in color. Native to Bolivia.

锦绣玉
Parodia comarapana

植株绿色，球形。10~14 条直棱，花靠顶生，鲜红色。原产南美洲玻利维亚和阿根廷海拔较高地区。

Body globular and green in color. 10 to 14 straight ribs. Flowers bright red and grow near the apex. Native to the high altitude areas of Bolivia and Argentina.

舞绣玉（快武丸）
Parodia comarapana

植株单生，5 cm 高，8 cm 宽。18~23 根周刺。中刺 3~4 根，略长于周刺，2 cm 长，黄色带有棕色尖。花蛋黄色，直径 0.5 cm。原产玻利维亚。

Solitary, up to 5 cm high, 8 cm wide. 18 to 23 radial spines. 3 to 4 central spines, relatively longer than radial spines, 2 cm long, yellow in color with brown tips. Flowers yolk yellow, 0.5 cm in diameter. Native to Bolivia.

暗绿锦绣玉
Parodia fuscato-viridis Bkbg.

植株单生，球形，约 4.5 cm 宽。花直径 3.5~6 cm，黄色。原产阿根廷北部。

Solitary, globular, up to 4.5 cm wide. Flowers are 3.5 to 6 cm in diameter, yellow in color. Native to northern Argentina.

丽花球
Parodia grandiflora

植株球形，绿色至碧绿色。具14~16条较浅直棱。刺座较密，长在棱背上。8~10枚周刺。花大鲜红色，花心较浅。原产南美洲玻利维亚等地。
Body globular, colors range from green to emerald. 14 to 16 shallow straight ribs. Areoles grow densely on the rib edges. 8 to 10 radial spines. Flowers big and bright red in color. Native to Bolivia and other areas in South America.

郝氏锦绣玉
Parodia hausteiniana Rausch

植株单生，短圆柱形，可达5 cm宽。花直径10 mm，黄色。原产玻利维亚。
Solitary, body short cylindrical in shape, up to 5 cm wide. Flowers 10 mm in diameter, yellow in color. Native to Bolivia.

胡梅尔锦绣玉
Parodia hummeliana Lau & Wesk.

植株单生，6 cm 高，7 cm 宽。原产阿根廷。
Solitary, 6 cm high and 7 cm wide. Native to Argentina.

卡拉锦绣玉
Parodia krahnii Wesk

植株单生或丛生，30 cm 高，8 cm 宽。花黄色。原产玻利维亚。
Solitary or clumping, up to 30 cm high and 8 cm wide. Flowers are yellow. Native to Bolivia.

劳氏锦绣玉
Parodia lauii Brandt

植株单生，可达 7 cm 高，9 cm 宽。花直径 4 cm，浅橙色至红色。原产玻利维亚。
Solitary, up to 7 cm high and 9 cm wide. Flowers 4 cm in diameter, colors range from salmon to red. Native to Bolivia.

马里锦绣玉
Parodia malyana Rausch

植株单生，高达 6 cm，宽 5 cm。中刺笔直。花直径 4 cm，黄色或红色。原产阿根廷。
Solitary, up to 6 cm high and 5 cm wide. Central spine straight. Flowers up to 4 cm in diameter, yellow or red in color. Native to Argentina.

梅赛德斯锦绣玉
Parodia mercedesiana Wesk.

植株单生，高达 5.5 cm，宽 6.5 cm。花直径 6 cm，红色或黄色。原产阿根廷北部。

Solitary, up to 5.5 cm high and 6.5 cm wide. Flowers are 6 cm in diameter, red or yellow in color. Native to north Argentina.

青王丸
Parodia ottonis (Lehm.) N.P. Taylor [*Notocactus ottonis*]

植株丛生。头部直径可达 11 cm。7~18 根周刺，0~4 根中刺，长约 2.5 cm。花黄色，直径 2~6 cm。原产巴西。

Clumping. Heads up to 11 cm high and wide. 7 to 18 radial spines, 0 to 4 central spines, up to 2.5 cm long. Flowers yellow in color, 2 to 6 cm in diameter. Native to Brazil.

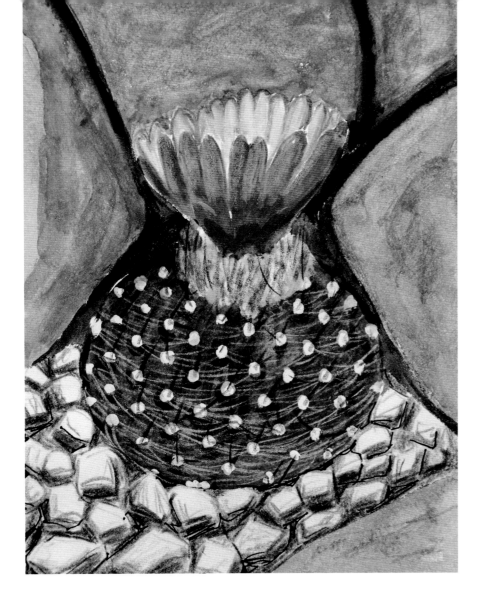

血红花锦绣玉
Parodia sanguiniflora Fric ex Bkbg.

植株单生，球形至长球形，宽大于 5 cm。花直径 4 cm。原产阿根廷北部。

Solitary, globular to elongated globular in shape, over 5 cm in width. Flowers 4 cm in diameter. Native to northern Argentina.

绯绣玉
Parodia sanguiniflora Fric. ex Bsvkeb.

植株球形，高可达 8 cm，具 14~16 条生有疣突的浅棱，周刺 15 枚，白色。中刺 4 枚。花顶生，花色多样，红、血红、黄红等，非常好看。原产阿根廷，现已被广泛引种栽培。

Body globular, up to 8 cm high. 14 to 16 shallow ribs with tubercles grow on the surface. 15 white radial spines and 4 central spines. Flowers terminal with various colors, such as red, blood red, yellowish red, etc. Native to Argentina, and widely cultivated all around the world.

红绣玉
Parodia schwebsiama (Werd.) Bkby.

植株球形，深绿色，长有12~14条厚直棱。周刺刚毛状，6~8枚，黄褐色，往下垂。花大鲜红色，漏斗状。原产阿根廷北部和玻利维亚。

Body globular and dark green. 12 to 14 thick straight ribs. 6 to 8 yellowish brown radial spines, bristle~like, bend downwards. Flowers big and bright red, funnel~shaped. Native to northern Argentina and Bolivia.

绯宝球（变种）
Paradia sanguiniflora var. comata Ritt.

斯图海锦绣玉
Parodia stuemeri (Werd.) Bkbg.

通常单生，高约 20 cm，宽 15 cm。25 根细周刺。花 4 cm 长，黄色至橙黄色。变种 var.robustior Bkbg. 具有更粗的刺，12~13 根周刺。花红色，直径 1~5 cm。原产阿根廷北部。

Usually solitary, up to 20 cm high, 15 cm wide. 25 radial spines, thin. Flowers 4 cm long, yellow to orange in color. var.robustior Bkbg. has thicker spines, 12 to 13 radial spines. Flowers are red, 2 to 5 cm in diameter. Native to northern Argentina.

韦伯锦绣玉
Parodia weberiana Brandt

植株单生，高达 7 cm，宽 10 cm。花黄色。原产阿根廷北部。

Solitary, up to 7 cm high and 10 cm wide. Flowers are yellow. Native northern Argentina.

精巧丸

Pelecyphora aselliformis C. Ehremb.

精巧殿

Pelecyphora aselliformis Ehrenbg.

植株球形，长满小斧头形灰白色疣突，相互紧密挤压在一起。刺呈篦齿状。花顶生红色。原产墨西哥。

　　Body globular, covered with white, tiny axe-like tubercles. Flowers terminal and red in color. Native to Mexico.

金凤龙
Pilosocereus chrysacanthus (F.A.C. Weber) Byles et G. D. Rwoley

植株乔木状，从基开始分枝，株高可达 5 m，茎亮绿色。生有 9~12 条尖锐浅棱，刺座生有密集短小金黄色或金褐色绒毛。花粉红色。原产墨西哥。

Tree like, branched from the base. Up to 5 m tall, stems bright green. 9 to 12 sharp, shallow ribs. Areoles are thickly covered with short, tiny, golden or golden brown tomentum. Flowers pink. Native to Mexico.

巴西青毛柱
Pilosoereus glaucochrous

植株多单生，柱形，直立生长。茎浅蓝色，高 4 m，粗 5~7 cm。具 6~10 条具有疣突的锐棱。周刺 9~12 枚，长 1.5 cm。中刺 3~4 枚，长 4~5 cm，黄色。花顶侧生，白色或红色，花柄底部着生有白色和褐色绒毛。原产巴西。

Usually solitary. Columnar and erect. Stems pale blue, up to 4 m high, 5 to 7 cm in diameter. 6 to 10 sharp ribs with tubercles. 9 to 12 radial spines, up to 1.5 cm long. 3 to 4 central spines, 4 to 5 cm long, yellow. Flowers lateral and grow near the top, yellow or red. The back of the pedicel is covered with white or brown tomentum. Native to Brazil.

舟乘团扇
Quiabentia verticillata

植株树形或灌木形，2~15 m 高，3 cm 宽。刺 7 cm 长。叶片椭圆形，长达 5cm，宽 2 cm。花淡红色，1.5 cm 长。原产玻利维亚。

Tree like or shrub like. 2 to 15 m high, 3 cm wide. Spines are 7 cm in length. Leaves oval in shape, up to 5 cm long, 2 cm wide. Flowers pale red in color, 1.5 cm in length. Native to Bolivia.

新玉（白银宝山）
Rebutia buiningiana

植株单生，亮绿色，5 cm 宽。14~16 根周刺，白色透明，0.6~1 cm 长。2~3 根中刺，1.4 cm 长。花粉色，直径 3 cm。原产阿根廷。

Solitary, light green in color, up to 5 cm wide. 14 to 16 radial spines, glassy white in color, 0.6 to 1 cm in length. 2 to 3 central spines, 1.4 cm long. Flowers pink, 3 cm in diameter. Native to Argentina.

眼球子孙球
Rebutia euanthema (Bkbg.) Buin. & Don.

原产阿根廷。
Native to Argentina.

华斯子孙球
Rebutia huasiensis Rausch

花直径约 3.5 cm。原产玻利维亚。
Flowers up to 3.5 cm in diameter. Native to Bolivia.

绯宝球
Rebutia krainziana Kesselr.

植株扁球形，深绿色具14~16条螺旋状的棱。刺座长于棱缘上，白色，周刺短小，8~12枚，花大红，花心黄色。长于球的基部。原产南美洲阿根廷。

Body flat globular, and dark green. 14 to 16 spiral~shaped ribs. Areoles grow on the rib edges, white. 8 to 12 short tiny radial spines. Flowers red, center yellow, grow at the base. Native to Argentina.

Rebutia kupperiana Boed.

花 3.5 cm 宽。原产玻利维亚。

Flowers 3.5 cm wide. Native to Bolivia.

白花子孙沟宝球
Rebutia leucanthema

头部短圆柱状，7 cm 高，3.5 cm 宽，黑绿色。7~8 根棕色周刺，长达 0.6 cm，紧紧贴在植株上。花白色，有时有粉色条纹，直径 2.5 cm。原产玻利维亚。

Heads short~cylindrical. 7 cm high and 3.5 cm wide, blackish green. 7 to 8 radial spines, brown in color, up to 0.6 cm long, attached tightly to the body. Flowers white, sometimes with pink stripes, 2.5 cm in diameter. Native to Bolivia.

白花沟宝球
Rebutia leucanthema Rausch

花直径 2.5 cm，白色或粉色。变种 var.paucicostatum Ritt. 较长，具更密集的刺，花猩红色。原产玻利维亚。

Flowers 2.5 cm in diameter, white or pink in color. var.paucicostatum Ritt. has longer body, denser spines, and scarlet flowers. Native to Bolivia.

Rebutia pygmaea (Fries) B. & R.

花直径 2.5 cm，亮红色至深紫红色。原产阿根廷北部。

Flowers 2.5 cm in diameter, bright red to dark purple-red in color. Native to north Argentina.

白宫丸变种
Rebutia pygmaea var. friedrichiana

植株体型大于其他变种。高约 5 cm，宽 3 cm。花直径 3 cm。原产玻利维亚。

Body is larger than other variants. Up to 5 cm high and 3 cm wide. Flowers 3 cm in diameter. Native to Bolivia.

Rebutia pygmaea var. haagei

花直径 4 cm。原产阿根廷 Jujuy。

Flowers 4 cm in diameter. Native to Jujuy, Argentina.

黑丽球
Rebutia rauschii

植株多群生，扁球形，株高 5 cm，株幅 10 cm。16 个以上呈螺旋状排列的矮疣突，黑绿色至紫色。白绒毡状刺座上，着生黄色或黑色周刺，花红色喇叭状。原产南美洲玻利维亚和阿根廷。

Usually clumping. Body flat-globular. Up to 5 cm tall, 10 cm wide. 16 short tubercles arrange in spiral shape, dark green or purple. Yellow or black radial spines grow on the white, fluffy areoles. Flowers red and trumpet~shaped. Native to Bolivia and Argentina.

茶柱
Stenocereus thurberi (Engelm.) Buxb.

植株基部能分枝，树型，多丛生。茎为深绿色。植株高 3~7 m，株幅 20 cm，12~19 条棱，周刺 7~10 枚，中刺 1~3 枚，刺褐色或灰黑色。花生于茎基部，粉红色带白花边，其果实可食用。原产墨西哥和美国西南部。

Tree like, usually clumping, branched at the bottom. Stems dark green. 3 to 7 m high, 20 cm wide. 12 to 19 ribs, 7 to 10 radial spins, 1 to 3 central spines, brown or greyish black. Flowers grow on the stem base, pink and with white margins. Fruits edible. Native to Mexico and southwestern U.S.

有沟宝山球
Sulcorebutia glomerispina (Card.) Buin. & Don.

原产玻利维亚 Huakani。
Native to Huakani, Bolivia.

黑丽丸
Sulcorebutia inflexiseta (Card.) Don.

分布于玻利维亚。
Distributed in Bolivia.

瓦斯奎斯有沟宝山球
Sulcorebutia losenickyana Rausch

植株单生,茎高达 7 cm。花鲜红色。原产玻利维亚。
Solitary, stems up to 7 cm tall. Flowers are bright red in color. Native to Bolivia.

梦托莎
Sulcorebutia santiaginiensis Rausch

植株体型较小。原产玻利维亚 Aiquile。

Body small. Native to Aiquile, Bolivia.

轮刺有沟宝山球
Sulcorebutia verticillacantha Ritt.

花淡紫色或朱红色带有橙色颈部。var.minima Pilbeam ex Rausch 变种茎小，花洋红色。原产玻利维亚。

Flowers pale purple or with an orange throat. var.minima Pilbeam ex Rausch has tiny stems and magenta flowers. Native to Bolivia.

大统领
Thelocactus bicolor (Gal. ex Pfeiff.) B. & R.

广泛分布于美国德克萨斯州西南部以及墨西哥北部。

Widely distributed in southwest Texas, and northern Mexico.

博拉大统领
Thelocactus bicolor var. bolaensis (Runge) Knuth

因其密布的淡黄白色刺而闻名。不同于原种的是，其花心并不是红色。原产墨西哥科阿韦拉州 Sierra Bola 地区。

Noted for its beautiful dense covering of pale yellowish white spines. Flowers lack the red center usually found in the species. Native to Sierra Bola, Coahuila, Mexico.

瓦兹大统领
Thelocactus bicolor var. schwarzii (Bkbg.) Anderson

多数无中刺。原产墨西哥 Tamaulipas, 以及 S of Llera 地区。
Central spine usually absent. Native to Tamaulipas and S of Llera, Mexico.

赤岭丸
Thelocactus hastifer (Werd. & Boed.) Knuth

植株单生，长圆柱形。花直径 3.5~5 cm。原产墨西哥的克雷塔罗州。
Stem solitary, long~cylindrical. Flowers 3.5~5 cm in diameter. Native to Queretaro, Mexico.

天晃
Thelocactus hexaedrophorus (Lem.) B. & R.

植株和刺的外形多变。花白色。var. lloydii (B. & R.) Kladiwa & Fittkau 变种具有非常明显的宽疣突以及粉色花朵。广泛原产墨西哥 San Luis Potosi, Tamaulipas, 以及 Nuevo Leon 等地区。

Spines and body shapes are variable. Flowers white. var. lloydii (B. & R.) Kladiwa & Fittkau has very conspicuous broad tubercles and pink flowers (Zacatecas, e.g., near Fresnillo). Native to San Luis Potosi, Tamaulipas, and Nuevo Leon, Mexico.

龙王丸
Thelocactus setispinus (Englm.) Anderson

植株单生，高达 12 cm，9 cm 宽。花直径 3~4.2 cm。原产美国德克萨斯州南部以及墨西哥。

Solitary, up to 12 cm high, 9 cm wide. Flowers 3~4.2 cm in diameter. Native to south Texas, and Mexico.

具棱瘤玉球
Thelocactus tulensis (Poselger) B. & R.

植株单生或丛生。花直径 3.5~4.2 cm，白色具淡粉色条纹。原产墨西哥。
Solitary or clumping. Flowers 3.5 to 4.2 cm in diameter, white with pale pink stripe. Native to Mexico.

具棱瘤玉球变种
Thelocactus tulensis var.matudae (Sanchez-Mej. & Lau) Anderson

植株单生，宽达 15 cm。花直径 7.5~8 cm。原产墨西哥。
Solitary, up to 15 cm wide. Flowers 7.5 to 8 cm in diameter. Native to Mexico.

金光龙
Trichocereus camarguensis

植株灌木状。茎高 50 cm，5 cm 宽。12~13 根周刺，3 cm 长。2~3 根中刺，5 cm 长。所有的刺均很细，针状。花白色，20 cm 长。原产玻利维亚。

Shrub like. Stems up to 50 cm high, 5 cm wide. 12 to 13 radial spines, 3 cm long. 2 to 3 central spines, 5 cm long. All spines are thin and needle~like. Flowers are white, 20 cm long. Native to Bolivia.

黑凤
Trichocereus coquimbanus

茎高达 1 m，宽 8 cm。从底部分枝。8~12 根周刺，1~2 cm 长。3~4 根中刺，5 cm 长。花白色，12 cm 长。原产智利。

Stems up to 1 m high, 8 cm wide. Branching from the base. 8 to 12 radial spines, 1~2 cm long. 3 to 4 central spines, 5 cm long. Flowers are white, 12 cm long. Native to Chile.

金棱
Trichocereus shaferi

植株从底部分枝。茎高 50 cm，宽 12 cm，鲜绿色。约有 10 根细刺，1.2 cm 长，无法区分周刺和中刺。花白色，长达 18 cm。原产南美洲阿根廷。

Branching from the base. Stems to about 50 cm high and 12 cm broad, fresh green. About 10 spines, thin, about 1.2 cm long. Radial and central spines are not distinguishable. Flowers white, up to 18 cm long. Native to Argentina.

利升龙
Trichocereus litoralis

植株直立或匍匐，从底部分枝。茎 2 m 高，12 cm 宽。15~29 根周刺，2~6 根中刺，2.4 cm 长。花白色，直径 10 cm。原产智利。

Body erect or sprawling, branching from the base. Stems up to 2 m high and 12 cm wide. 15 to 29 radial spines, 2 to 6 central spines, 2.4 cm long. Flowers are white, 10 cm in diameter. Native to Chile.

皎丽玉
Turbinicarps lophophoroides Buxb. Et. Bckbg.

颜色白绿。菱形瘤底下，尖端长出短刺 4~5 根。顶部特别多绵毛，花白色。
Body whitish green, 4 to 5 short spines growing under and at the tip of the tubercles. Tomentum grow on top. Flowers white in color.

精巧殿
Turbinicarpus pseudopectinatus (Bkbg.) Glass & Foster

植株单生，5 cm 高，2~3.5 cm 宽。var.rubriflora 变种具有粉红色花。原产墨西哥 Tamaulipas 以及 Nuevo Leon 地区。

Solitary, up to 5 cm high and 2~3.5 cm wide. var.rubriflora has reddish pink flowers. Native to Tamaulipas and Nuevo Leon, Mexico.

金刺尤伯球
Uebelmannia flavispina

Body globular, 10 to 12 cm high, 10 to 12 wide, greyish grenn in color. Stems have 15 to 18 ribs, areoles densely growing on the ribs, white and bright yellow tomentum grow on the areoles. Flowers funnel~shaped, yellow in color, 2 to 3 cm in diameter. Appear in summer.

原产巴西东部。植株球形。株高 10~12 cm，株幅 10~12 cm。茎具 15~18 条棱，表皮灰绿色，棱上刺座排列密集，刺座着生白色绒毛和鲜黄色刺，整齐排列在棱上。花漏斗状，黄色，花径 2~3 cm。花期夏季。

类节球
Uebelmannia pectilera var. pseudopectinfera

植株球形或圆筒形，株高 30~50 cm，株幅 15 cm。茎具有 12~18 条棱，翠绿色，刺座生有较密的绒毛，刺红褐色，生长交错。花漏斗形，黄色。原产巴西。

Body globular or cylindrical. 30 to 50 cm tall, 15 cm wide, 12 to 18 ribs, emerald. Thick tomentum grow on the areoles. Spines reddish brown. Flowers yellow, funnel~shaped. Native to Brazil.

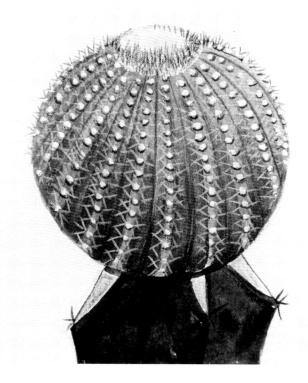

黄刺尤伯球
Uebelmannia flavispina Buin. & Bred.

植株单生，高达 35 cm，宽达 11 cm。花直径 7 mm，淡黄色。原产巴西 Minas Geraes 地区。

Solitary, up to 35 cm high, 11 cm wide. Flowers 7 mm in diameter, light yellow. Native to Minas Geraes, Brazil.

黄刺节齿尤伯球
Uebelmannia pectinifera Buin.

植株单生，高达 50 cm，宽达 15 cm。花黄绿色。var.pseudopectinifera Buin. 变种植株更小更绿，刺向周边伸展。原产巴西 Minas Geraes 地区。

Solitary, up to 50 cm high, 15 cm wide. Flowers greenish yellow. var. pseudopectinifera Buin. is smaller, body greener, spines spreading sideways. Native to Minas Geraes, Brazil.

花细球

Weingartia neumanniane (Backeb.) Werclerm.

植株扁球形或球形，由 12~14 个扁平疣突构成。刺座较大，生有白色粉状绒毛。周刺 6~8 根，黑针状刚刺，刺长 4~6 cm，花大顶生，金黄色，漏斗形。原产玻利维亚。

Body flat globular or globular, and is made of 12 to 14 flat tubercles. Areoles relatively large, covered with white tomentum. 6 to 8 black, needle~like radial spines. Flowers big and golden, funnel~shaped, terminal. Native Bolivia.

百合科 (Liliaceae)

百合科近期与其它科合并后成为一个庞大家族，约有 250 属 3700 多种，其肉质类植物主要分布在芦荟属 (Aloe)、十二卷属 (Haworthia) 和鲨鱼掌属 (Gasteria)，其中以芦荟属最为丰富，包括杂交栽培种超过了 500 种，很多种属由于形态、色泽多姿多彩，已成为家居庭院和植物公园的常见观赏花卉，芦荟属有些种类成为药物、保健品和化妆品的重要原料，被大量栽种。

After combining with other families, the Liliaceae become a large family, includes about 250 genera and 3700 species. Succulent plants of this family are distributed mainly in the genera Aloe, Haworthia and Gasteria. There are more than 500 species, including hybrid cultivars, in genus Aloe, most of them are colorful and diversified in morphology. They become common ornamentals in private courtyards and plant parks. Some plants of genus Aloe are widely cultivated due to their medicinal and nourishing values..

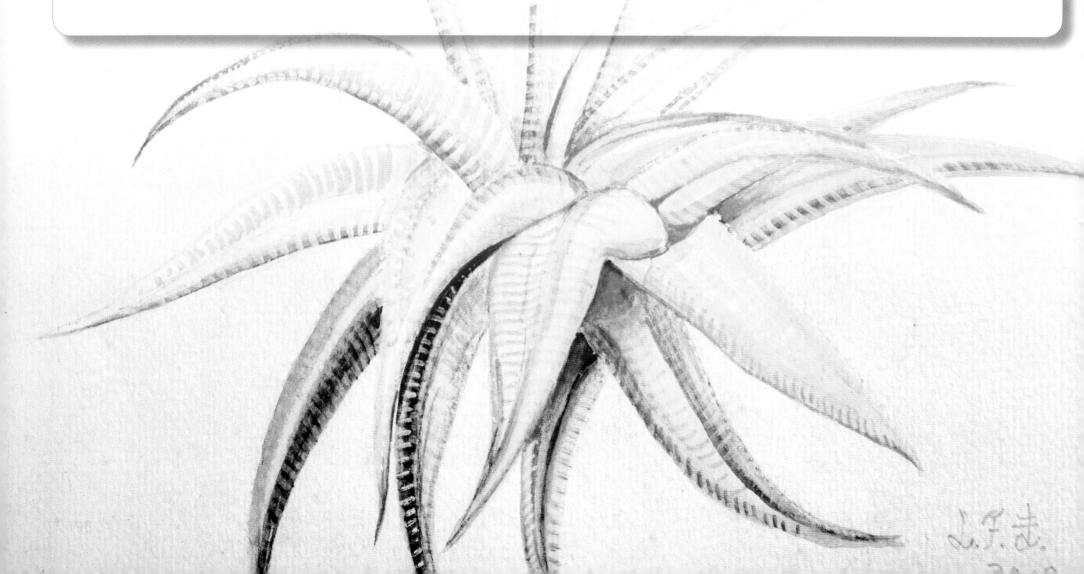

铃丽锦（皮刺芦荟）
Aloe aculeata

单生，叶长而阔，两面均有许多从白色小突起上长出的皮刺。花序是由许多花密集排成。直立，长圆柱形，花蕾橙红色，开花后转黄。分布于南非北部、津巴布韦等开阔的稀树草原或草原上的石砾地上，可供观赏。

Single, leaves long and broad, numerous prickles arising from white tubercles found on both surfaces. Inflorescence erect, long cylilndric, densely crowded by many flowers. Flower-buds orange and turned yellow in blossom. Distributed in northern South Africa and Zimbabwe. Found in the open bush-veld or gravel ground in veld. Cuttivated for ornament.

非洲芦荟
Aloe africana

茎细长，高约2~4 m，叶上下表面均有数颗小刺。花序单生，每株可抽出数支。花蕾向下，开花后上翘，橙黄色。分布于南非的南部杂树丛中。

The stem slender , up to 2~4m, a few spines on both surfaces of leaves. Several in florescences arised in one plant. Flowers downward in bud stage but upward in blossom, orange in cluor. Distributed in bush-wood of the southeast part of South Africa.

盎格鲁芦荟
Aloe angelica

茎细长，高 4~5 m，叶狭长，平展或下弯。花序呈球形，花蕾红色，开花后转黄。生长于南非北部茂密的灌木丛或开阔地带。其叶汁化学成分与木本芦荟相似。也可作药用。

Stem slender, 4~5m high; leaves narrow, horizonal or recurved. Inflorescence spheroidal. Flowers red in bud stage and yellow in blossom. Distributed in dense bush-wood or open ground. The chemical composition of the leaf-sap similar to that of Aloe. Arborescens and used for medicine too.

布尔芦荟
Aloe bubrii

成熟植株可分裂为多个莲座状叶丛，叶黄绿带红色，两面均有许多"H"形白色斑点，花序多分枝，花黄色，也有橙红色。分布于南非西部一小范围，喜生于少雨干旱地区。

Mature plants divided into several rosettes, leaves yellowish-green mixed with red, numerous H-shape white sports on both surfaces, inflorescence multi-branched, flowers yellow inclined to orange-red. Distvibufed in a low small area of western South Africa.

绿花芦荟
Aloe chlarantha

丛生，叶绿色，有时转暗红至紫红色，叶两面有小白点，枯叶留茎。花序高达 3 m，不分枝，长圆柱形。花小（约 1 cm）黄绿色，花蕾隐藏于大苞片内，开花时才伸出来。分布于南非中部偏南内陆的石山北坡。

Tufted, leaves green, turned dull red or purplish red in drought, white spots on both surfaces, the wirher leaves left, inflorescence up to 3m. Unbranched, long cylindric, flowers small, about 1cm, yellowish-green, flower-buds hidden in large bracts, spread out blossom. Distributed in a small area of the south part of the northern slope of rocky hill.

箭筒芦荟（二分叉芦荟）
Aloe dichotoma

主茎粗壮，树皮有皱纹，枝条多为二分叉，花序直立，多分枝，花黄色。喜生于砂石地，分部在南非的中西部地区，当地人用它的空心枝条做箭筒用，故得名。

Stout trunk with creased bark, two-forked branches, inflorescence erect, much-branched, flowers yellow. Distributed in the sandy soil areas of central and western parts of South Africa.

The hollow branches are used for quiver making by local people and so named "Quiver tree".

津巴布韦芦荟（高芦荟）
Aloe excelsa

茎单生，枯叶留茎，花序直立多分枝，可多达数十个圆柱形花序，花红色。喜生于温暖干旱的山谷或小山岗上。分布于南非北端，莫桑比克，津巴布韦，赞比亚和马拉维等地。

Stem single with withered leaves, inflorescence erect, multibranched, sometimes more than ten, cylindric, flowers red. Found in warm, dry walleys or hills. Distributed in northern South Africa, Moxambique, Zimbabwe, Zambia and Malawi.

格斯特芦荟
Aloe gerstneri

单生，叶肉质，表面光滑，花序单生，成熟时偶有分枝，长圆锥形，花淡橙色。分布于南非中部偏东一小地区的石头坡上，为稀有种。

Single, leaves fleshy, glossy, inflorescence single, occasionally branched in mature, long conical, flowers light orange. Distributed in a small area of eactern central South Africa, found on rocky slope. A strange species of Aloe.

鬼切芦荟
Aloe mailothii

巨箭筒芦荟
Aloe pissansii

高大乔木，花序多分枝，花黄色。分布于南非西部及纳米比亚。为南非著名的特色植物。

A tall tree, inflorescence multibranched, flowers yellow. Distributed in western South Africa and Namibia. A famous featured plant of South Africa.

超雪芦荟
Aloe rauhii' Super Sanow Flake'

单生，叶肉质，表面光滑，花序单生，成熟时偶有分枝，长圆锥形，花淡橙色。分布于南非中部偏东一小地区的石头坡上，为稀有种。

Single, leaves fleshy, glossy, inflorescence single, occasionally branched in mature, long conical, flowers light orange. Distributed in a small area of eactern central South Africa, found on rocky slope. A strange spicies of Aloe.

雷诺兹芦荟
Aloe reynoldsii

丛生，叶阔，折曲成波浪形，淡蓝绿至黄绿色，两面均有"H"形的白点并连成纵长线。叶缘有粉红色细线，具细齿。花序多分枝，花黄色，顶部绿色。生于南非东部 Bashee 河口的潮湿悬崖峭壁上。

Tufted, leaves broad, twisted as wavy, pale bluish-green to yellowish-green, with H-shape white spots and linked up as longitudinal lines on both surfaces, margin, denticulate and with pink line, inflorescence multi-branched, flowers yellow with green top. Distributed in the Bashee River mouth of eastern South Africa, found on cliffs.

艳丽芦荟（歪头芦荟）
Aloe speciosa (Tit-head)

叶片不规则地排列成莲座，着生于茎的一侧，叶蓝绿色，有一窄红线在边缘围绕，具细齿。花序梗短，呈圆柱状的花序基部藏于叶丛中，花蕾红色，开花后转绿白色，红色的雄蕊伸出花外，常成群生长。耐中等霜冻和抗虫害。分布于南非南部，是美丽的观赏植物。

A beautiful ornamental plant, leaves irregularly arranged as a rosette on one side of the stem, the blade bluish-green with red edge, serrulate, peduncle short, the base of culindric raceme hidden in the rosette, flower bud red, turned to greenish-white in blossom, the red stamens spread out of the corolla.

Growing in groups, frost-resistant and insect-tesistant, distributed in south part of South Africa.

奇丽芦荟
Aloe spectabilis

单茎，高大，可达 4 米以上。花序粗壮，多分枝。花密集排成圆柱形花序，直立，鲜橙红色。分布于南非 Kwazulu-Natal 省。

Stem single up to 4m or higher. Inforescence strong, multibranched, flwers densely arranged as cylindric raceme, erect, bright orange. Disributed in South Africa.

瓦奥姆比芦荟
Aloe vaombe

单茎，叶平展，花序多分枝，长圆锥形，花红色。分布于马达加斯加岛。

Stem single, leaves flat, inflrescence multi-branched, long conical, flowers red. Distributed in Madagascar island.

恐龙卧牛（短叶卧牛一种）栽培种
Gasteria var. (Hybrid)

霸龙
Gasteria cvs.

卧牛锦
Gasteria nitida var. Armstrongii 'Variegata'

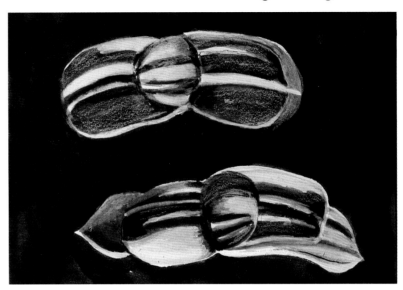

比兰龙锦
Gasteria 'Variegata'

卧牛龙白覆轮
Gasteria×Gagyu-nishki

白摺鹤
Haworthia marginata

为一较大型盆栽观叶花卉,肉质叶较坚实,原产非洲南部及西南部,现已被广泛引种各地区。

A relatively huge potted ornamental; leaves succulent and solid; native to south and southwest Africa; widely introduced all around the world.

点纹冬之星座
Haworthia pumila

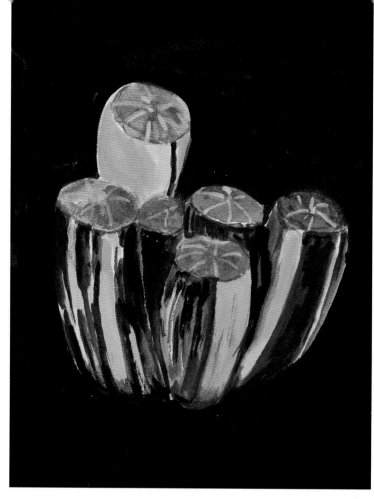

万象锦

Haworthia truncata 'Variegata'

玉扇锦

Haworthia truncata 'Variegata'

鬼瓦
Haworthia 'Onigawara'

景天科 (Crassulaceae)

景天科约有 30 多个属 1500 余种，为多年生低矮灌木，有些为藤本，是肉质植物重要的科属之一。分布较为广泛，在温暖、干燥地区都能看见它们。叶对生、互生和或轮生，绝大部分肉质化，为此其生长形态多样，色泽变化鲜艳，为花卉爱好者所喜欢，现已被广泛引种到世界各地，是盆栽观叶植物的重要成员。

The plants of family Crassulaceae, include approximately 1500 species and 30 genera, are generally perennial low shrubs but sometimes vine. They are one of the major family of succulent plants, and are extensively distributed in both warm and/or dry areas. Their leaves are opposite, alternate or verticillate, and mostly succulent. Due to the diversity of their growth morphology and bright color, the plants of family Crassulaceae are deeply favored by flower and plant enthusiasts and widely introduced all around the world, which makes them an important member of potted ornamentals.

黑法师
Crassula arboreum (Hybrid)

姬星（黑法师变种）
Crassula nepextirs 'Tom Thumb'(Hybrid)

新花月锦
Crassula ovata 'Variegata'

艳镜
Masonia deperssa

番杏科 (Aizoaceae)

番杏科植物虽然在我国无自然分布，但其奇特的形态十分具有观赏价值，已经大量为我国植物公园和爱好者引种、栽种。番杏科原产于南非和纳米比亚，具有独特生态环境，干旱、昼夜温差大、半沙漠和岩石裂缝中。全科共有 100 多属，已经鉴定的有 2000 多个种。该科为多年生肉质草本或矮小灌木，花单生雏菊形。在我国不宜露地栽种，只能在具有温床和温度调节的温室内栽种。

Although the plants of genus Aizoaceae are not naturally distributed in China, they are widely introduced by plants enthusiasts and parks due to their particular morphologies. They are native to South Africa and Namibia and grow in unique environment, such as rock cracks and semi-desert. About 100 genera and 2000 species in this family. Mostly perennial succulent herbs or small shrubs; flowers solitary, daisy like. They are unsuitable for outdoor cultivation in China and could only be cultivated in greenhouse.

慈光锦
Cheridopsis demiculata

翔凤
Cheridopsis pecaliaris

奇凤玉
Dinteranthus microspernis

玉帝
Plciospilos ndlii

结束语

一年多前一本《600种手绘多肉植物图谱》出版后，在清理残存一些有关照片和资料时，发现尚有不少还未被搬出来使用，很可惜，为此，在多位同窗同事鼓励和帮助下，决心整理400余种以仙人掌科植物为主的有关肉质植物素材，再请夏念和教授、林侨生工程师审定名字和我的老搭档吴萍副研究员、王辰博士研究生合作完成全部资料。重要的是我以前的学生徐金富先生和程世法先生，他们两位现已是非常成功人士，全力赞助出版所需费用，在此感谢再感谢！对张贺先生在资料运送和传递方面做了不少工作表示感谢，出版社王斌先生和他的团队为这400余种手绘多肉植物图谱续集作出的努力，在此也一一致谢。

朱亮锋等

A year ago, my book 《600 Colour Paintings of Succulent Plants》, containing 600 species of succulent hand-painted manuscripts, was published. Whilst tiding up my desk, remnants of drawings, photos and manuscripts were found and each was just waiting to be collated into a further botanical document. With the encouragements and assistance from a number of my colleagues and scholars, I determined to bring those remaining 400 manuscripts into light. They describe a large number of Succulent and Cactus plant species. Engineer, Mr Xia Nianhe and Professor Lin Qiao-sheng have audited, identified and authorized the botanical names on each page. With the collaboration from my much respected associate Wu Ping and Dr. Wang Chen, I was able to collate all these artifacts to be published.

I am sincerely grateful to have the sponsorship from my previous students Mr Xu Jin Fu and Mr Cheng Shi Fa, both of whom are now successful professionals. Without their help, the publishing requirements would not have been possible. Thank You from the bottom of my heart! I also like to express my sincere gratitude to Mr Zhang He for his thoroughly work on the data collection and transmission.

My appreciation also goes to Mr Wang Bin and his team in the publishing agency for the tireless work and collaborative effects to make this sequel 《400 Colour Paintings of Succulent Plants》 to become a published reality. Thank you all!

Zhu Liang Feng

拉丁名索引
Index to Scientific Names

A
Acanthocalycium thionanthum 3
Acanthocalycium thionanthum var. aurantiacum 3
Aloe aculeata 197
Aloe africana 197
Aloe angelica 198
Aloe bubrii 198
Aloe chlarantha 199
Aloe dichotoma 199
Aloe excelsa 200
Aloe gerstneri 200
Aloe mailothii 201
Aloe pissansii 201
Aloe rauhii 202
Aloe reynoldsii 202
Aloe speciosa 203
Aloe spectabilis 203
Aloe vaombe 204
Ariocarpus agavoides 4
Ariocarpus furfuraceus 4
Arrpjadpa penicillata. 5
Arthroereus glaziovii 5
Astrophytum asterias 'Caespitosa' 6
Astrophytum asterias 'Nuolas Variegaia' .. 6
Astrophytum asterias 7
Astrophytum asterias × A. Myriostigma ... 7
Astrophytum capricorne 8
Astrophytum myriostigma 8
Astrophytum myriostigma 9
Astrophytum myriostigma var. 'Hexagenus' 10
Astrophytum myriostigma var. 'Nudum' 11
Astrophytum myriostigma var. quadricostatum 'Nuclum' 10
Astrophytum myriostigma var. quadricostatum 9
Astrophytum mytiostigma var. columuare 'Heptagonus' 11
Astrophytum ornatum 12
Astrophytum ornatum var. bubescente .. 13
Astrophytum senile 13

B
Bergerocactus emoryi 14
Borzicactus roezlii 15
Borzicactus samnensis 15
Borzicactus samnensis 16
Borzicactus sericatus 16
Brouningia hertlingiana 17

C
Cacti sp. 17
Cereus chalybaeus 18
Cereus jamacaru 'Variegata' 18
Cereus peruvianus 19
Chamaecereus silvestrii var. aurea 19
Chamaelobivia Senrei-Gyoku 'Variegata'. 20
Cheridopsis demiculata 215
Cheridopsis pecaliaris 215
Cleistoactus smarageliflora 20
Cleistocactus ritteri 21
Copiapoa cinered var. 21
Copiapoa grandifiore 22
Copiapoa hypogaca 22
Copiapoa lauii 23
Copiapoa tenuissima 24
Copiapoa tenuissima 23
Copinapoa cinerea var. columna-alba.... 24
Corphantha calipensis 26
Corryocactus anrens 25
Coryphantha bumamma 26
Coryphantha cornifera 27
Coryphantha echinus 27
Coryphantha gracilis 28
Coryphantha lauii 28
Coryphantha longicornis 29
Coryphantha recurvata 29
Coryphantha robust 30
Coryphantha robustispina 30
Coryphantha salm-dyckiana 31
Coryphantha sulcata 31
Coryphantha sulcolanata 32
Crassula arboreum 211
Crassula nepextirs 212
Crassula ovata 213
Cylindropuntia prolifera 32
Cylindropuntia versicdor 33

D
Denmoza erythrocephala 34
Denmoza rhodacantha 34
Denmoza erythrocephala 33
Dinteranthus microspernis 216
Discocactus heptacathus. 35
Discocactus horstii 35
Discocactus insignis 36
Disocacta nelsonii 36
Disocacta phyllanthoides. 37

E
Echinocactus grusonii 'Breris-niruis' 38
Echinocactus grusonii 'Tansirinshuehi' 38
Echinocactus grusonii 37
Echinocactus horizonthalonius 39
Echinocactus ingens 39
Echinocactus platyacatus 40
Echinocactus texensis 'Hybride' 40
Echinocactus texensis 'Hybride' 41
Echinocactus texensis 41
Echinocereus adustus 42
Echinocereus adustus var. schwarzii 42
Echinocereus barthelowanus 43
Echinocereus berlandieri 43
Echinocereus bristolii var. pseudopectinatus 44
Echinocereus chisoensis 44
Echinocereus dasyacanthus 45
Echinocereus engelmannii 45
Echinocereus engelmannii var. nicholii.. 46
Echinocereus enneacanthus 46
Echinocereus enneacanthus-Hybrida..... 47
Echinocereus fendleri 47
Echinocereus fendleri 48
Echinocereus ferreirianus 48
Echinocereus ferreirianus var. lindsayi 49
Echinocereus grandis 49
Echinocereus Hybrida 50
Echinocereus johnsoni 50
Echinocereus knippelianus 51
Echinocereus lauii 51
Echinocereus leucanthus 52
Echinocereus longisetus 52
Echinocereus ochoterenae 53
Echinocereus palmeri. 53
Echinocereus pamanesiorum 54
Echinocereus papillosus 54
Echinocereus pectinatus 55
Echinocereus pectinatus var. 55
Echinocereus polyacanthus var. densus 56
Echinocereus pulchellus. 56
Echinocereus pulchellus var. amoenus ... 57
Echinocereus reichenbachii var. baileyi 57

Echinocereus reichenbachii var. baileyi 58
Echinocereus reichenbachii var. fitchii... 59
Echinocereus schmollii 59
Echinocereus sciurus var. floresii............ 60
Echinocereus sp. 60
Echinocereus spinigemmatus 61
Echinocereus stoloniferus subsp. tayopensis
.... 61
Echinocereus stramineus 62
Echinocereus subinermis var. ochoterenae
.... 62
Echinocereus triglochidiatus var. melana-canthus-Hybrida.................................. 63
Echinocereus triglochidiatus var. mojavensis
.... 64
Echinocereus triglochidiatus var. neomexicanus 64
Echinocereus triglochidiatus 63
Echinocereus viereckii 65
Echinocereus viridiflorus......................... 65
Echinocereus viridiflorus var. davisii 66
Echinoereus triglochidiatus...................... 66
Echinomastus warnockii 67
Echinocereus 'Hybrida'........................... 67
Echinopsis ancistrophora 68
Echinopsis camdicans 69
Echinopsis cardenasiana 69
Echinopsis eyriesii................................... 70
Echinopsis eyriesii var. (Hybride)............ 71
Echinopsis haematantha........................... 71
Echinopsis hertrichiana............................ 72
Echinopsis mamillosa.............................. 73
Echinopsis minuana................................. 73
Echinopsis molesta 74
Echinopsis obrepanda var. calorubra 74
Echinopsis smrziana 75
Echinopsis sp. 75
Echinopsis tortispina 76
Echinopsis-Hybrida (Red)....................... 72
Echinopsis-Hybrida (Red)....................... 76
Epiphyllum chrysocardium...................... 77
Eriosyce ausseliana 78
Eriosyce susseliana.................................. 78
Escobaria leei... 79
Escobaria minima 79
Escobaria roseana 80
Escobaria vivipara 80
Espostoa melanostele 81
Espostoopsis dybowskii 81
Eulychnia acida 82
Eulychnia saint-pieana 82

F

Ferocactus acanthodes............................. 83
Fercocactus echidne................................. 83
Ferocactus echidne.................................. 84
Ferocactus emoryi.................................... 84
Ferocactus flavovirens 85
Ferocactus fordii...................................... 85
Ferocactus hamatacanthus 86
Ferocactus hamatacanthus 87
Ferocactus hamatacanthus 86
Ferocactus hamatacanthus var. hamatiden
.... 87
Ferocactus histrix 88
Ferocactus latispinus................................ 88
Ferocactus macrodiscas 89
Ferocactus schwarzii................................ 89
Frailea asteriocks 91
Frailea chiquitema 90
Frailea horstii 90

G

Gasteria var. (Hybrid) 204
Gasteria cvs.. 205
Gasteria nitida var. Armstrongii' Variegata'
.... 205
Gasteria' Variegata' 205
Gasteria×Gagyu-nishki 206
Ggmnocactus buenekeri............................ 91
Gylmnocalycium mihnovichu var. friedrichii 'Rubra'... 92
Gymnoalycium saglione 'Variegata'........ 92
Gymnoalycium tillanum........................... 93
Gymnocactus baldianum........................... 94
Gymnocactus gielsdorfianus 94
Gymnocalycium megalothelos 95
Gymnocalycium achirasense.................... 96
Gymnocalycium ambatoense 96
Gymnocalycium andreae........................... 97

Gymnocalycium anisitsii. 97
Gymnocalycium baldianum var. 98
Gymnocalycium bodenderianm var. F. Verieg.. 98
Gymnocalycium bruchii 99
Gymnocalycium buenekeri 99
Gymnocalycium capillaense.................. 100
Gymnocalycium carminanthum 101
Gymnocalycium chiquitamum 101
Gymnocalycium damsii 102
Gymnocalycium denuclatum-Hybrida (Red) 102
Gymnocalycium denudatum 103
Gymnocalycium denudatum-Hybrida . 103
Gymnocalycium eurypleurum 104
Gymnocalycium ferrarii 95
Gymnocalycium fleischerianum 104
Gymnocalycium gibbosum 105
Gymnocalycium griseopallidum 105
Gymnocalycium heischerianum 106
Gymnocalycium horstii 106
Gymnocalycium hyptiacathum 107
Gymnocalycium leeanum 107
Gymnocalycium mihanovichii var. 'Variegata' 108
Gymnocalycium mihanovichii var. friedrichii 109
Gymnocalycium mihanovichii var. friedrishii 108
Gymnocalycium nigriareolatum 109
Gymnocalycium ragonesii..................... 110
Gymnocalycium ritterianum................. 110
Gymnocalycium schatzlianum 111
Gymnocalycium schickendantzii 111
Gymnocalycium schroederianum van ....
.... 112
Gymnocalycium sp. 114
Gymnocalycium sp. 112
Gymnocalycium sp. 113
Gymnocalycium sp. 114
Gymnocalycium spegazzinii..................115
Gymnocalycium stellatum 93
Gymnocalycium tudae var. izozogsii 115
Gymnocalycium tudae var. pseudomala-cocarpus 116
Gymnocalycium uebelmannianum........ 116
Gymnocalycium weissianum................. 117
Gymnocalycium horridispinum............. 117

H

Harrisia martini 118
Haworthia marginata 207
Haworthia pumila 207
Haworthia truncata 208
Haworthia 'Onigawara'......................... 209
Hamatocactus uncinatus 118

I

Islaya krainzina..................................... 119

L

Lobivia arachnacantha.......................... 119
Lobivia arachnacantha var. densiseta 120
Lobivia aurea var. shaferi..................... 121
Lobivia aurea 120
Lobivia bruchii 121
Lobivia caineana 122
Lobivia cardenasiana 122
Lobivia cinnabarina 123
Lobivia famatimensis 123
Lobivia famatiminsis 124
Lobivia grandiflora var. crassicaulis 124
Lobivia haematantha var. amblayensis ... 125
Lobivia haematantha var. hualfinensis ... 126
Lobivia haematantha var. rebutioides ... 126
Lobivia haematantha............................. 125
Lobivia jajoiana 127
Lobivia kieslingii 128
Lobivia rauschii 128
Lobivia rosariana 129
Lobivia saltensis 129
Lobivia schieliana................................. 130
Lobivia schreiteri var. stilowiana 130
Lobivia silvestrii 131
Lobivia sp. 131
Lobivia tiegeliana var. cinnabarina 132
Lobivia wrightiana var. winteriana...... 132
Lobivia zecheri 133
Lophephora wieeianmsii 133
Lophophora cliffusa.............................. 134

Lophophora diffusa 135

M

Malacocarpus erinacaus................... 135
Mammillaria albiflora 136
Mammillaria boolii 137
Mammillaria chiocephala.................. 137
Maimmillaria conspicua.................... 136
Mammillaria giselae 138
Mammillaria napina 138
Mammillaria pottsii 139
Mammillaria spinosisimna................. 139
Mammillaria tetrancistra.................... 140
Mammillaria wrightii 140
Masonia deperssa 213
Matucana crinifera 141
Matucana intertexta.......................... 141
Matucana pujupatii 142
Melocactus guaricensis 142
Melocactus sp. 143
Melocactus violaceus......................... 144
Micranthocereus flaviflorus 144
Myrtillocaotus geomertrzaus 145

N

Nopalxochia ackermanii 149
Neobuxbaumia euphorbroide 145
Neochilenia paucicostatus 146
Neollogolia conoidea var. 146
Neoporteria heinrichiana 147
Neoporteria napina 147
Neoporteria occulta.......................... 148
Neoporteria subgibbosa var. vastanea.... 148
Noeporteria brachyacanthus.............. 149
Notocactus caespitosus...................... 150
Notocactus crassigibbus.................... 151
Notocactus crassigibus...................... 158
Notocactus fuscus............................. 151

Notocactus herteri............................. 152
Notocactus magnificus..................... 152
Notocactus magnificus..................... 153
Notocactus mueller-melchersii.......... 153
Notocactus muricatus 154
Notocactus ottonis 154
Notocactus roseiflorus...................... 155
Notocactus roseolueus var. 155
Notocactus seopa var....................... 150
Notocactus tephraeanthus 156
Notocactus uebelmannianus............. 156
Notocactus veenianus 157
Notocactus warasii 157

O

Oputia azurea 163
Opuntia durangensis....................... 158
Opuntia ficus-Indica 159
Opuntia gosseliniana 159
Opuntia microdasys var. albispina 160
Opuntia monacantha 160
Opuntia phaeacantha....................... 161
Opuntia phaecantha var. camanchia.... 161
Opuntia pheacantha 162
Opuntia subututa 162
Oreocereus hendriksenianus............. 163

P

Parodia aureicentra 164
Parodia aureispina 165
Parodia chrysacanthion.................... 165
Parodia columnaris......................... 166
Parodia comarapana........................ 166
Parodia comarapana........................ 167
Parodia fuscato-viridis..................... 167
Parodia grandiflora......................... 168
Parodia hausteiniana 168
Parodia hummeliana....................... 169

Parodia krahnii 169
Parodia lauii 170
Parodia malyana 170
Parodia mercedesiana 171
Parodia ottonis................................ 171
Pachycereus pringlei........................ 164
Parodia sanguiniflora...................... 172
Paradia sanguiniflora var. comata.... 173
Parodia schwebsiama...................... 173
Parodia stuemeri 174
Parodia weberiana 174
Pelecyphora aselliformis................... 175
Pilosocereus chrysacanthus 176
Pilosoereus glaucochrous 176
Plciospilos ndlii............................... 216

Q

Quiabentia verticillata 177

R

Rebutia buiningiana........................ 177
Rebutia euanthema 178
Rebutia huasiensis 178
Rebutia krainziana.......................... 179
Rebutia kupperiana 179
Rebutia leucanthema 180
Rebutia pygmaea............................. 181
Rebutia pygmaea var. friedrichiana..... 181
Rebutia pygmaea var. haagei 182
Rebutia rauschii 182

S

Stenocereus thurberi........................ 183
Sulcorebutia glomerispina................ 183
Sulcorebutia inflexiseta.................... 184
Sulcorebutia losenickyana................ 184
Sulcorebutia santiaginiensis 185
Sulcorebutia verticillacantha 185

T

Thelocactus bicolor.......................... 186
Thelocactus bicolor var. bolaensis 186
Thelocactus bicolor var. schwarzii........ 187
Thelocactus hastifer......................... 187
Thelocactus hexaedrophorus 188
Thelocactus setispinus 188
Thelocactus tulensis 189
Thelocactus tulensis var.matudae........ 189
Trichocereus camarguensis 190
Trichocereus coquimbanus 190
Trichocereus litoralis 191
Trichocereus shaferi 191
Turbinicarps lophophoroides............ 192
Turbinicarpus pseudopectinatus 192

U

Uebelmannia flavispina 194
Uebelmannia flavispina 193
Uebelmannia pectilera var. pseudopec-
tinfera 193
Uebelmannia pectinifera 194

W

Weingartia neumanniane 195